abc's of
ELECTRONICS

by
Farl J. Waters

Howard W. Sams & Co., Inc.
4300 WEST 62ND ST. INDIANAPOLIS, INDIANA 46268 USA

International Standard Book Number: 0-672-21507-1
Library of Congress Catalog Card Number: 77-93167

Printed in the United States of America.

Preface

This third edition of *abc's of Electronics* is a simple, easy-to-follow text devoted to the fundamentals of electronics. Every possible effort has been made to avoid technical concepts and mathematical terms. Presentation of the subject is enhanced by the use of simple language and analogies familiar to everyone.

Included in this text are discussions of atomic structure and *current carriers*—the electrons and the holes—and the parts they play in electricity. The electric and magnetic forces causing current, and the various forms of opposition and effects encountered by that current are covered.

Because transistors are being used more and more in electronic equipment, they are given extensive coverage in this edition. Both the construction and operating principles are covered. Numerous illustrations make the text easy to follow. Circuit action of transistors and other solid-state devices is explained. Principles and construction of integrated circuits (ICs) are explained and illustrated.

Later chapters give basic theory on radio waves and how they are produced. Basic oscillator circuits using transistors are covered.

At the end of each chapter are a number of questions that will test the knowledge gained from your reading. Answers to these questions are given in an appendix. If you find that you have not retained the correct answers, a review of that chapter is advisable.

It is my sincere hope that this book will give you a basic knowledge of electronics and a desire to acquire more knowledge of the subject.

FARL J. WATERS

Contents

CHAPTER 1

The Electron

The word electricity began with application of the Greek term *elektron* by Dr. William Gilbert, M.D. (English 1540-1603). From the Greek word elektron, meaning amber, came the term electrics. In 640 B.C., a Greek by the name of Thales found that static electricity could be produced by briskly rubbing a rod of amber. Later, the American, Benjamin Franklin (1706-1790), proved lightning to be electricity. Both the scientist and the layman considered this unseen thing a fluid.

It was not until the late 1890's that the English scientist Sir Joseph John Thomson and others ushered in the Electronic Age by explaining atomic structure. The *electron theory* that evolved from that discovery defined electricity as the movement or the accumulation of electrons. Thus, the common conception of electronics being a part of electricity has been falsely acquired by both layman and scientist. The term electronics is commonly assigned to that portion of the science making use of the vacuum tube (the radio tube) or the newer solid-state devices such as transistors, silicon controlled rectifiers, variacs, triacs, etc. The term electric is generally used with regard to the development of power, light, heat and similar applications.

ATOMIC STRUCTURE

All matter is composed of small units called *molecules.* The molecule is so small that it is difficult to give a comparison for its size except to note that the smallest particle of matter one can imagine may contain many thousands of molecules. In turn, the even smaller

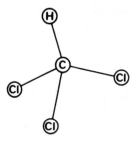

Fig. 1-1. Relationship among the 5
atoms of a chloroform molecule.

atom is a subdivision of the molecule. The number of atoms, their arrangement, and the type or types of atoms which compose the molecule determine its character. The character of the molecules within a material identifies that material. For instance, rubber and paraffin both contain carbon and hydrogen atoms; however, the rubber molecule has many thousands of carbon atoms as compared with the paraffin molecule which has 25 to 30 atoms of carbon. Chloroform molecules (Fig. 1-1) have the shape of a three-sided pyramid constructed of five atoms—one of hydrogen, one of carbon, and three of chlorine. As another example, atoms of copper combine with other atoms of copper into molecules to form the metal which is recognized by its distinctive color as being copper.

Fundamentally, all matter is composed from combinations of 104 different atoms or elements. As might be expected, each element has atoms of differing character. Within the atom is a nucleus made up primarily of *protons* and *neutrons* surrounded by electrons revolving in a number of orbits. Fig. 1-2 shows the complexity of the uranium atom as compared with that of a hydrogen atom. The latter has only one proton circled by one electron while the uranium atom has 92 protons, 143 neutrons, and 92 electrons.

Within the nucleus of the atom the neutrons and protons are bound together by the most powerful force known. Even though this nuclear force binding the protons and neutrons together is now being used for many purposes, including the powering of U.S. Navy submarines, very little is known of its true nature. It is said that these nuclear forces are a million times stronger than those holding the oxygen atoms and the hydrogen atoms together within a molecule of water.

TYPES OF ELECTRICITY

Extending the experiments of Thales the Greek, Benjamin Franklin found that rubbing a glass rod produced a form of electricity differing from that produced by rubbing a rubber rod. It was also noted that when a charge developed on the glass rod was transferred to two lightweight pith balls, it caused them to repel each other.

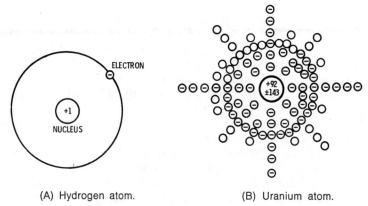

(A) Hydrogen atom.   (B) Uranium atom.

Fig. 1-2. Comparison between atoms.

However, the two pith balls were attracted when one was charged from the glass rod and the other from the one made of hard rubber. Since the two forms of electricity appeared to be opposite, Franklin labeled one as positive and the other as negative. Actually the assignment of positive or negative to either of the two forms of electricity was a matter of choice. At that particular time Franklin may have supposed the positive electricity nearest his goal.

Later it was discovered that negative electricity is associated with an excess of electrons and the definition of a negatively charged particle was assigned to the electron. Conversely, positive electricity is then a deficiency of electrons or an excess of positively charged protons. Within the normal atoms, the number of electrons is equal to the number of protons, hence a neutral charge condition exists. Since the protons are so tightly bound within the nucleus, only the electrons are free to create the excess or deficient condition referred to as electricity. Thus, in the study of electronics we are primarily interested in the activity of electrons. However, as we shall learn there are within the confines of crystalline materials positively charged particles referred to as *holes*. Consequently, electrons and holes have been labeled *current carriers* and their movements are referred to as *current*. But since most current is outside crystalline materials and is the movement or flow of electrons, the terms current and electron flow are used somewhat interchangeably.

ELECTRON MOVEMENTS

Normally, the movement of electrons is in the form of orbits around the nucleus of the atom. Within the hydrogen atom, one electron moves about the nucleus in a single orbit or shell. But seven orbits are required for the 92 electrons in the uranium atom.

9

Since each orbit must be spaced away from the adjacent orbit, the force exerted by the protons to hold the electron within an orbit varies with each orbit. Consequently, those electrons beyond the second orbit are more or less free to move from one orbit to another orbit or from atom to atom. Accordingly, the copper atom has 19 free electrons that can move from an original orbit to another orbit or from the original atom to another.

Theory states that a small amount of this electron movement occurs continually in all matter. But if this movement can be concentrated so that 6,280,000,000,000,000,000 electrons pass a given point in one second, it can produce enough heat to raise the temperature of 0.00093 pound of water by 1° F. At first this might seem to be an enormous number of moving electrons required to accomplish such a small feat. However, within a one inch length of No. 14 B & S copper wire there are approximately 70,000,000,000,000,000,-000,000 free electrons (about 12,000 times the number of electrons considered above). Heat created by movement of electrons is the result of the friction between those electrons as they jostle about from one atom to the next.

An electron is somewhat analogous to a drop of water. Alone, neither the drop of water nor the electron can exert much force. But in years past many drops of water turned the giant mill wheel and today the same milling operation is performed by the proper channeling of many millions of electrons. In the same sense that the water of the stream has to be brought to bear its force on the mill wheel, the flow of electrons must follow the windings of the motor before the wheat can be ground into flour.

Thus, by knowing that either the drop of water or the electron will seek its neutral level through the path of least resistance, man can make both work for him. It is easy to see how a drop of water makes its way through the soil, ditches, streams, and rivers to the sea—its point of neutral level. But it is a different world—a world of smallness, extreme complexity, and, to some extent, abstraction—through which the electron may travel to enter the neutral level of an atom. Of course, the atom is only neutral when its number of electrons equals that of its protons. Any unbalance (electric charge) is neutralized only when free electrons seeking to produce this neutral condition unite with that atom lacking its normal number of electrons.

REVIEW QUESTIONS

1. Who was Thales and what did he discover? GREEK — Stat. E/e
2. What is the Greek word meaning amber? ElektRoN
3. What is the movement or the accumulation of electrons called?
   zlectRicity (electRoN TheoRy)

4. Name the small unit of which all matter is composed and its smaller subdivision. MOLECULES — ATOM
5. How many differing atoms exist? 104
6. What kind of atoms are found in copper? In rubber? In chloroform? COPPER   CARBON & HYDROGEN — HYDROGEN CARBON CHLORINE
7. Name the three particles found within the atom.
8. What happens to the pith balls when charged with like charges? When charged with unlike charges?
9. Is the electron a positively or negatively charged particle?
10. What are free electrons?

7 PROTONS, NEUTRONS, ELECTRONS

8.

9. AN ELECTRON CAN BE + — CHARGED PARTICLE

10. FREE ELECTRONS ARE THOSE ELECTRONS TRAVELING IN ORBIT WHICH ARE PART OF AN ATOM BUT DO NOT TRAVEL AS CLOSE TO THE NUCLEUS AS OTHERS.

CHAPTER 2

Electricity and Magnetism

It is an old vaudeville gag that the cow does not give her milk, you have to take it from her. In the same sense, the atom does not give up its electrons, but force must be used to take the electron from its neutral position within the atom.

STATIC ELECTRICITY

Static electricity (electricity at rest) is the storage of positive or negative charges on a body. When a rubber rod is rubbed with a piece of fur, electrons taken from the fur are deposited on the rod. The rubber rod then has a negative charge—a charge that can be transferred to another object by touch. If the object to which the charge is transferred is quite large or not separated from other material by an *insulator*, the excess electrons will become scattered and lose their effectiveness. The insulator referred to is any material having an atomic construction that does not readily allow electron movement. Of the opposite characteristic is the conductor—any material that passes electrons easily. There is neither an absolute insulator nor an absolute conductor.

The electricity produced on the rubber rod is a matter of friction. Electricity produced by friction is commonly a static form used primarily in laboratories for research. In more common occurrences of static electricity, it is a nuisance and often damaging rather than useful. The printing industry uses extreme caution to reduce the static electricity which accumulates on paper rolling through the printing press. Common manifestations of frictional static electricity are when hair is attracted to a comb, lint to clothing, or when a spark jumps from one person to another.

Lightning is one form of static electricity that cannot be entirely attributed to friction. Usually during warm weather, clouds acquire a static charge that moves from cloud to cloud or from cloud to earth creating the spectacular flash we know as lightning. This visible flash with its accompanying thunder is the result of a great number of electrons forcing their way through the atmosphere, which is normally an insulator and does not allow electrons to flow through it.

DYNAMIC ELECTRICITY

Of greater importance and usefulness than static electricity is that form of electricity which has a continuous source of free electrons that can be drawn off, or discharged, at a chosen rate. This form of electricity is referred to as *dynamic* electricity.

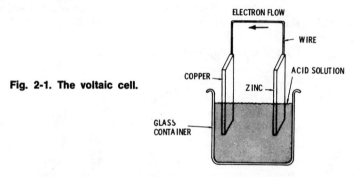

Fig. 2-1. The voltaic cell.

One method of producing dynamic electricity is provided through chemistry. In 1880, an Italian, Count Alessandro Volta, placed zinc and copper strips in an acid solution. When the two metal strips were connected by a conductor, as in Fig. 2-1, gas bubbles formed about the copper. Volta also observed that the zinc strip was being eaten away. The sulfuric acid solution he used contained positive hydrogen ions and negative sulfate ions. The *ion* is an atom or group of atoms having gained or lost electrons by combining with another atom or group of atoms. In forming sulfuric acid, two hydrogen atoms give up their two electrons to the sulfate group. The sulfate group thus becomes a negative ion that tends to combine with and lose its excess of electrons to the zinc strip. The electrons freed at the zinc strip then travel through the conductor to the copper strip where they are attracted to the positive hydrogen ion. As the hydrogen ion regains its electron, it is again a free atom of hydrogen which can escape into the air. If Volta could have had Edison's electric lamp connected between the copper strip and the zinc strip, the flow of electrons could have lighted the lamp.

This cell—the Voltaic Cell—was limited in its usage by the fact that the sulfate ions, the hydrogen ions, and the zinc are consumed in the process. However, it was this principle that led to our present dry cell (flashlight battery) and to the lead-acid battery used in our automobiles.

MAGNETIC ENERGY

The English physicist Michael Faraday (1791-1867) spent most of his life attempting to find a link between magnetism and electricity. As yet, that link has not been firmly established. However, we do now have a theory that gives us a means of studying the magnetic induction which causes electron movement. Loadstone or magnetite, a form of iron ore discovered near the city of Magnesia in Asia Minor about 585 B.C., was found to have the ability to attract pieces of iron and some other metals. These pieces of loadstone, or magnetite, are now referred to as magnets. Modern magnets are made of steel which has been treated to acquire the properties of a magnet. They are formed into a bar (Fig. 2-2A) or a horseshoe (Fig. 2-2B).

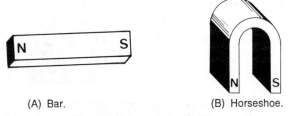

(A) Bar.   (B) Horseshoe.

Fig. 2-2. Permanent magnets.

When freely suspended in a horizontal position, the bar magnet will tend to swing until one end points in a northerly direction. It is then convenient to refer to that end, or pole, of the magnet pointing north as the North or N pole. It follows that the other end of the magnet is the South or S pole. Much like the positive and negative charges of electricity, like magnetic poles (N or S) repel each other. Conversely, unlike poles attract each other. Thus it appears that near the geographic North pole there is an area having the characteristics of the S magnetic pole. Because of this a lightweight freely moving magnet can be used as a compass for guidance along unmarked trails.

To study some characteristics of the magnet, place a bar magnet beneath a glass plate, and sprinkle iron filings on top of the glass.

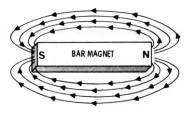

Fig. 2-3. Magnetic lines of force around a permanent magnet.

You will notice that the iron filings fall into definite lines about the magnet. These lines, which are referred to as *magnetic lines of force,* extend from one pole to the opposite pole, as seen in Fig. 2-3. The bunching of the iron filings indicates that the energy of the magnet exists in these magnetic lines of force. By pushing a piece of copper wire (which is nonmagnetic) under the glass plate into the magnetic lines of force, it is seen that the iron filings alter their position (Fig. 2-4). Much like the rubber band, these lines tend to give against the movement of the copper wire and are therefore elastic. This elasticity is limited, however, and there is a point at which the line will be broken or cut by the nonmagnetic material.

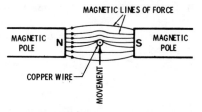

Fig. 2-4. Effect on the magnetic lines of force when a nonmagnetic conductor is passed through a magnetic field.

After the limit of elasticity has been exceeded and a magnetic line of force cut, this line will reform to its original condition. But as the nonmagnetic material (the copper wire) moves, the lines bunch up ahead of the movement, thus, concentrating more lines of force at the leading side of the copper wire. If the movement stops, the lines of magnetic force will reform to give equal concentration on each side of the nonmagnetic conductor.

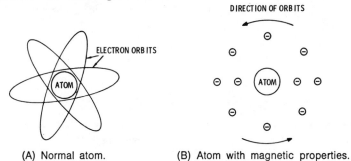

(A) Normal atom. (B) Atom with magnetic properties.

Fig. 2-5. Orbital paths of electrons around atoms.

15

The orbits of the electrons around an atom are normally of many planes, as shown in Fig. 2-5A. However, the present theory is that an atom of iron having magnetic properties has electron orbits all lying in one plane. All electrons are moving in the same direction within those orbits (Fig. 2-5B). The magnetic atom is then like the blades of a fan whirling about its hub. The forces of repulsion existing between the electrons of adjacent atoms will cause other atoms to align in much the same manner. Across the pole face of a bar magnet, the atoms form a pattern similar to that shown in Fig. 2-6. The movement of electrons along the outer edge of the pole is in one direction, as indicated by the large arrow.

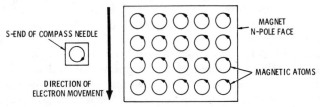

Fig. 2-6. Pattern formed by atoms across the N-pole face of a bar magnet.

A compass needle can be used to verify that the pole face shown in Fig. 2-6 is an N pole face. It would seem normal that an S-pole face would have an electron movement opposite that of the N pole, so that when the compass needle lines up with the large bar magnet the electron movement along adjacent edges would be in the same direction. Using the forefinger of your left hand to point in the direction of the magnetic lines of force, from N to S, note that the thumb indicates the direction of electron movement. Inversely, with the thumb of the left hand pointed in the direction of electron movement, the forefinger indicates the direction of the magnetic force.

A similar electron movement can be established within a wire. And if the magnetic needle of a compass is placed directly above the wire as in Fig. 2-7, the electron flow will cause that needle to come to rest at right angles to the wire. Thus, the movement of

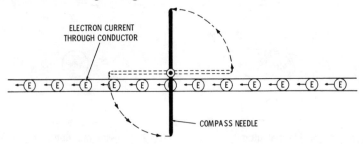

Fig. 2-7. The effect of electron flow on the magnetic needle of a compass.

electrons within the wire has produced lines of magnetic force circling that wire.

Magnetic Induction

Exchanging the cause for the effect is often found to be the answer to a complex problem. In our present study this would involve the use of magnetic lines of force to cause electron movement. One manner of forming such lines of force about an electron conductor (for example a copper wire) would be that shown in Fig. 2-8. In moving this copper wire (which is nonmagnetic) between an N pole and an S pole, the lines of magnetic force tend to bend and to

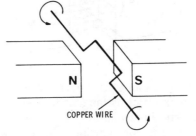

Fig. 2-8. Method of inducing current in a nonmagnetic conductor.

COPPER WIRE

concentrate ahead of the movement. As these lines bend they partially circle the wire causing a movement of electrons within it. This is known as *magnetic induction,* or the induction of an electron movement by magnetic forces. It must be noted that this induction is the result of three forces—magnetic, mechanical, and electromotive.

The three forces involved in magnetic induction are shown by the vectors (arrows) in Fig. 2-9. Accordingly, these three forces are considered to have a relationship represented by a three dimensional pattern with each force being perpendicular to the other two. Fol-

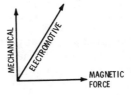

Fig. 2-9. The three forces associated with magnetic induction.

MECHANICAL
ELECTROMOTIVE
MAGNETIC FORCE

lowing the rule of reversed cause and effect, any two of these forces acting in the proper relationship can produce the third. For example, a magnetic force in proper relation with electron flow through a conductor can create a movement of the conductor as a result of the mechanical force developed. This is the principle by which the radio speaker converts changes in electrical energy into corresponding mechanical variations.

ALTERNATING WAVE DEVELOPMENT

As the nonmagnetic conductor in Fig. 2-10 is rotated between the two magnetic poles, it crosses, or cuts, the magnetic lines of force in one direction and then in the other, as indicated by the arrows. The resultant current within the conductor also changes directions as the conductor is rotated.

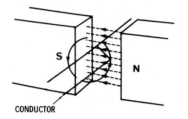

Fig. 2-10. Current is induced in this nonmagnetic conductor as it is rotated through the fixed magnetic field.

Beginning its movement at position P1 in Fig. 2-11, the conductor's action is tangent to the circular path and parallel to the magnetic lines of force. Therefore, at position P1 the conductor is not crossing or "cutting" the magnetic lines of force and no current results. At position P2 the movement is partially parallel and partially perpendicular to the magnetic lines. Since only that movement which is perpendicular to the lines of magnetic force produces current flow, the amount of electron flow produced at position P2 will

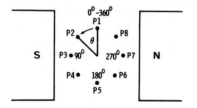

Fig. 2-11. Reference points for the rotating conductor in Fig. 2-10.

not be as great as that at position P3 where all movement is perpendicular. At position P3 the conductor is cutting a maximum number of magnetic lines per unit time and thereby inducing maximum current within the conductor. The perpendicular movement of the conductor at position P4 is the same as that at P2 and the current produced at each of these positions is equal. Likewise, the movement at position P5 is parallel to the magnetic lines of force and current in the conductor is zero just as it was at P1. During the second half of the revolution, through positions P6, P7, P8 and back to P1, the conductor crosses the magnetic lines of force in the opposite direction to produce an electron flow in reverse to that during the first half-cycle.

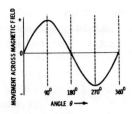

(A) Movement of the conductor
through the magnetic field.

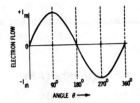

(B) Current induced in the conductor.

Fig. 2-12. The action of Fig. 2-11 plotted as sine waves.

The conductor movement across the magnetic lines of force can be expressed mathematically as being related to the trigonometric sine function:

$$p = A(\sin \theta)$$

And the instantaneous current values and instantaneous emf values are:

$$i = I(\sin \theta)$$
$$e = E(\sin \theta)$$

where,

    A, I, and E are the maximum values of effective movement, current, and voltage,

    θ is the angle between position P1 and the position being considered.

Fig. 2-12A shows the instantaneous variations of the conductor movements, while Fig. 2-12B indicates the variations of the current. As the conductor in Fig. 2-10 rotates from position P1 (Fig. 2-11) through a complete revolution, or cycle, the current increases from zero to a maximum in one direction, falls back to a zero value, then increases to maximum in the opposite direction and again falls to zero. The direction of current is commonly assigned positive or negative values. Thus, one half of the revolution, or cycle, is said to be positive while the reverse half-cycle is negative.

As the conductor is turned through successive cycles, the current continues to alternate. Hence, an *alternating electron flow* or alternating current (ac) results.

FREQUENCY

A certain amount of time is required for a cycle of alternating current. *Frequency* is the term denoting the number of cycles occurring in one unit of time (generally one second). Thus, if 0.01 second is required for the electron flow to complete one cycle of alternation, it will complete 100 cycles in one second which is a frequency of 100 cycles-per-second (cps). The term hertz (Hz) has been adopted to signify cycles per second. Thus, the unit of 100 hertz is the same as 100 cycles per second.

The ac brought into our homes and places of business alternates at 60 Hz.

Audio Frequency (AF)

When ac at frequencies between 20 and 20,000 Hz causes a diaphragm to vibrate, the resulting disturbances of the air create sound. Frequencies with this range are therefore known as the sound or *audio frequencies*. The speaker of a radio or a television set simply converts variations of current into corresponding mechanical movements. This, in turn, produces sound. With a microphone this process is reversed; mechanical variations are converted into audio.

Radio Frequency (RF)

Frequencies above 200,000 Hz are termed *radio frequencies* (rf). By feeding ac at such frequencies into a conductor isolated from the earth and having no normal return path to the current source, a form of electromagnetic radiation known as radio waves is produced. These waves, which travel through space at the speed of light, induce a similar current in any conductor they contact. It is this radiation which makes radio, television, and similar forms of communication possible.

ANALYSIS OF A SINE WAVE

Ac at a frequency of 100 Hz will make a complete cycle in 0.01 second. At the end of a complete cycle the angle θ will equal 360 degrees (or a multiple of 360 degrees as in Fig. 2-11). During 0.002 second a 100-Hz electron flow will complete 0.2 cycle which corresponds to angle θ of 72 degrees (0.2 × 360).

It becomes rather complex to carry this analysis through a complete cycle and the related mathematics is beyond the level of our study. However, the results of such an analysis are shown in Fig. 2-13. It should be noted that with a frequency other than 100 Hz, the time base—the horizontal scale—would be altered. Further illustra-

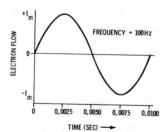

Fig. 2-13. The instantaneous values of electron flow at a frequency of 100 Hz, versus time.

Table 2-1. Squared Values of Instantaneous Current

Angle θ	Cycles	Current (I)	Current Squared (I²)
0	0	0	0
10	0.028	0.174	0.03
20	0.056	0.342	0.12
30	0.084	0.500	0.25
40	0.111	0.643	0.42
50	0.139	0.766	0.59
60	0.167	0.866	0.75
70	0.194	0.940	0.88
80	0.222	0.985	0.97
90	0.250	1.000	1.00
100	0.278	0.985	0.97
110	0.306	0.940	0.88
120	0.333	0.866	0.75
130	0.361	0.766	0.59
140	0.389	0.643	0.42
150	0.417	0.500	0.25
160	0.445	0.342	0.12
170	0.472	0.174	0.03
180	0.500	0.	0
190	0.527	—0.174	0.03
200	0.555	—0.342	0.12
210	0.583	—0.500	0.250
220	0.611	—0.643	0.42
230	0.639	—0.766	0.59
240	0.667	—0.866	0.75
250	0.695	—0.940	0.88
260	0.722	—0.985	0.97
270	0.750	—1.000	1.00
280	0.777	—0.985	0.97
290	0.805	—0.940	0.88
300	0.833	—0.866	0.75
310	0.859	—0.766	0.59
320	0.880	—0.643	0.42
330	0.917	—0.500	0.25
340	0.945	—0.342	0.12
350	0.973	—0.174	0.03
360	1.000	0	0.

Total 20.04
Mean 0.50
Root Mean 0.707

tion of the 100-Hz current variation can be seen by the instantaneous values given in Table 2-1.

RMS Values

In Fig. 2-14, the number of electrons passing through resistance R during the first half cycle of alternating current is equal to the

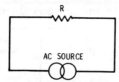

Fig. 2-14. A resistive ac circuit.

number of electrons traveling through R during the second half cycle. This would seem much like a football team in the first half of the game scoring 21 points and during the second half allowing their opponent to score 21 points. Regardless of which direction the current flows, a certain amount of work is being performed. The power which is absorbed as current passes through resistance R equals:

$$I^2R$$

where,

    I is the effective rate of current in amperes,
    R is the resistance in ohms.

Table 2-1 gives the squared values of the instantaneous current. Notice that these values are always positive. Determining the average, or mean, of these squared values, the *effective value* of current is found to be 0.707 of the maximum value. The effective value, 0.707 Im, is the second *root* of the *mean* of the *squared*-instantaneous values of current or the root-mean-square (rms) value. The effective value of the alternating electromotive force is similarly 0.707 of its maximum value. Unless otherwise stated, specifications dealing with alternating current or emf will be expressed as an rms value. Meters for measuring alternating emf generally indicate rms values.

THE DC GENERATOR

Fig. 2-15 illustrates the basic operating principle of a dc generator. The ends of the rotating conductor are arranged to make contact with two semicircular metal plates. As the end of conductor A is in contact with plate 1, the current flow is toward plate 1. When the end of conductor B makes contact with plate 1 the current flow

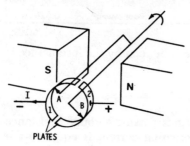

Fig. 2-15. A basic dc generator.

is reversed (now leaving contact B) but still flowing toward plate 1. Thus, plate 1 is always negative. Plates 1 and 2 form what is known as a *commutator*. Since the electron flow from the commutator does not alternate, it is a form of direct current—actually a *pulsating direct current* (Fig. 2-16). By using a commutator comprised of many conductors spaced about the center of rotation and insulated

Fig. 2-16. Pulsating direct current.

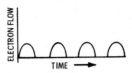

from each other, an almost pure direct current can be obtained. Now, as the current from one conductor begins to fall off, that of another conductor is rising to a maximum value, thereby producing an almost constant rate of current.

THE AC GENERATOR

An ac generator (Fig. 2-17), on the other hand, having three conductors rotating about a common center has three different outputs. When rotating conductor A is producing 0.866 of its maximum current, conductor B is just ready to begin its cycle and at this instant is producing no current. In other words A has completed ⅓ of a cycle when conductor B is ready to start the cycle. The current

Fig. 2-17. A basic ac generator.

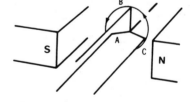

of conductor B is said to *lag* and to be *out of phase* with conductor A by ⅓ cycle (120°). Conductor C produces a current that *lags* the electron flow variations produced by conductor B. However, the alternations produced by conductor C are ahead of, or leading, those of conductor A by 120°. Since the current induced in each of the three conductors is out of phase with the others, such a generator is referred to as a *3-phase power* source. Power can also be generated in 2-phase, 4-phase, etc. Wherever electrical motors of greater than one horsepower are required, the use of 3-phase current is normal. Lines transmitting power from a generating station to an area where the power is to be consumed usually supply

3-phase ac. Occasionally 6-phase power is supplied; however, this can readily be converted to 3-phase.

TRANSFORMER ACTION

When an alternating voltage is applied to an inductance the magnetic field has a similar alternating characteristic. That alternating magnetic field induces an opposing and equal voltage into the winding P of Fig. 2-18. If the applied alternating voltage is 100 volts and the number of turns in winding P equals 100, the induced

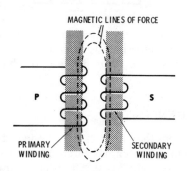

Fig. 2-18. Principles of transformer action.

voltage is 1.0 volt per turn. With the same alternating magnetic field acting upon a second or secondary winding S, a similar voltage will also be induced into each of its turns. At a rate of 1.0 volt per turn the voltage induced into a secondary winding S of 50 turns equals 50 volts. Thus, the voltages across the primary winding P and the secondary winding S are directly proportional to their number of turns.

$$\frac{E_p}{E_s} = \frac{N_p}{N_s}$$

where,

E_p is the voltage applied across the primary winding,
E_s is the voltage induced into the secondary winding,
N_p is the number of turns in the primary winding,
N_s is the number of turns in the secondary winding.

The magnitude of a magnetic field is proportional to the current and the number of turns within a winding. Accordingly, the magnetic field developed by the primary winding of Fig. 2-18 is proportional to its current (I_p) and its number of turns (N_p). The relative magnetic field, (θ) is the product of the current and number of turns of the primary. Inversely, it is this same magnetic field that causes the secondary current (I_s) in the secondary turns (N_s). Therefore:

$$\theta = I_s N_s$$
$$\text{and } I_p N_p = I_s N_s$$
$$\text{or } I_p / I_s = N_s N_p$$

Thus, if our 100 primary turns have a current of 1.0 ampere, the 50 secondary turns will have a current of 2.0 amperes. A device like that shown by Fig. 2-18 is known as a *transformer*. Its transformer action is that of transforming a high voltage or current into a lower voltage or current or that of transforming low voltage or current to higher values. Transformers can be used to step up the 120-volt ac to 300 volts for use with vacuum tubes or to step down that same 120 volts ac to 25 volts or less for transistor usage. Impedances are also stepped up or down by the transformer action.

REVIEW QUESTIONS

1. What is electricity at rest called?
2. What is an insulator? A conductor?
3. What form of electricity is produced by friction?
4. What is lightning?
5. What is the term given to that form of electricity that moves steadily?
6. Is there a definite link between magnetism and electricity?
7. Is it the N or the S pole of the magnet that points to the geographic North pole?
8. In what form does the force of the magnet exist?
9. Describe how a copper wire alters and cuts magnetic lines of force.
10. If the electrons are moving about a magnetic pole face in a clockwise manner, is it an N or an S pole?
11. What is magnetic induction?
12. Can a mechanical force be developed through the use of magnetic forces and electron flow?
13. Describe the variations in the rate of current flow produced as a conductor rotates in a magnetic field.
14. What is frequency?
15. What is the frequency of the current produced by a battery?
16. What is the frequency of the current used in our homes?
17. What are audio frequencies? Radio frequencies?
18. What is a commutator?
19. What is 3-phase power?
20. Describe a very basic form of the transformer. If a primary has 100 turns and an applied voltage of 100 volts, what is the voltage across the secondary of 360 turns?

CHAPTER 3

Impedance to Current Flow

Just as highway traffic speeds are limited by many factors, electrical current flow is limited or impeded in a number of ways.

RESISTORS AND CONDUCTORS

In a manner similar to the force of the exploding power in the rifle, an electromotive force (abbreviate emf) can cause a free electron to move into the midst of a medium of atoms and their electrons. If the medium into which a free electron is caused to move has an atomic structure which yields freely to that force, the movement of the electrons can be quite rapid. Such metals as copper, silver, aluminum, and zinc are among those mediums which have atomic structures that permit rapid movement of free electrons.

In contrast, the small closely fitted atoms of carbon have electrons that are tightly held to the nucleus and present a considerable resistance to the movement of electrons. Thus, carbon is a poor conductor of electric current. *Resistor* is the term given to mediums (such as carbon) that resist, or impede, the flow of electrons. On the other hand, materials such as silver, copper, aluminum, or zinc that present little resistance to electron flow are known as *conductors*. Materials such as glass, mica, and rubber provide greater resistance to the flow of electrons than carbon and are therefore termed *insulators*.

Although an insulator presents a high resistance to electron flow, there is no material that will not pass electrons when enough force is applied. Dry air is commonly considered an insulator; however, a force of 50 volts can cause electrons to pass through 0.001 inch of dry air. This passage of electrons through dry air is accomplished only if a great deal of energy is used—energy that can be seen and heard as an electrical spark. When the voltage is extremely high, it is capable of traveling a great distance through air—for example, a flash of lightning.

The opposite of resistance is *conductance*—the ability of material to pass electrons. A resistor has high resistance and hence low conductance, while zero conductance would be the characteristic of a perfect insulator.

ELECTRICAL UNITS OF MEASUREMENT

Electronics, like any other science, would be nothing without measurements and numbers. It was a German electrician, G. S. Ohm, who first established in 1827 a relationship for mathematic reasoning in electronics. Ohm determined that any increase in emf would cause a proportional increase in current. Therefore, doubling the emf will double the flow of current. The ratio of the emf to the resulting current flow is a constant. *E* is used to represent electromotive force and *I* the current. As long as the path of the current is not altered, its E/I ratio will remain a constant value. If the constant value of the E/I ratio is 2, an emf of 4 units can cause a current of 2 units, and 4 units of current would be effected when 8 units of emf are applied.

Altering the path of electron movement by adding resistance to give an E/I ratio of 4 will allow a current of 1 unit when acted on by an emf of 4 units. The resistance of the circuit or electron path is therefore the factor determining the E/I ratio. If a 1-inch length of carbon rod with a uniform diameter has an E/I ratio of 2, a similar rod 2 inches in length would have an E/I ratio of 4. It would be normal to assume that a carbon rod of twice the length would allow for a passage of electrons at half the rate of the shorter carbon rod. Resistance is therefore in direct relation with the E/I ratio, and the numeric value of E/I can be taken as a measure of the resistance:

$$R = \frac{E}{I}$$

where,

E is the electromotive force,
I is the current in amperes,
R is the resistance in ohms.

This is referred to as Ohm's law.

Electromotive Force

The electromotive force, or electrical pressure, which causes current to flow is measured in units called *volts*. The emf is commonly referred to as voltage. Here both terms will be used. Similarly the flow of electrons has been labeled as *current*. Actually, the term current originated even before the electron theory when it was first thought that there was some sort of flow associated with electricity.

PRESS = Volts

The Ampere and the Coulomb   *GPM = AMP*

The unit for measuring current is the *ampere*, derived from the name of a French physicist, André Marie Ampere (1775-1836). By legal definition an ampere of electron flow passing through a solution containing silver ions will deposit 0.001118 gram of silver in one second. One ampere is also defined as the flow of 6,280,000,000,-000,000,000 electrons passing a given point in one second of time, which is the same as the movement of one *coulomb* in one second. Fortunately, a person need not be burdened with these definitions; he can depend on meters designed to indicate the amount of current directly in amperes, milliamperes, etc. Similarly, emf values can be measured with voltmeters.

To better understand how electrons flow, consider the two buckets of water shown in Fig. 3-1. The bucket on the left represents a negative potential which has an excess of electrons. The one on the right represents a positive potential which has an electron deficiency. Both buckets are connected by a small pipe with a shutoff valve in the middle. The difference in the water level of the two buckets represents the potential difference, or voltage. If the valve were to be opened, the excess water in the negative bucket would flow into the positive bucket until the level in both were the same. This is similar to the action that occurs when a conductor is connected across a source of potential difference. Electrons always flow from negative to positive.

Electrical Power

As a quantity of electrons pass between points having a potential difference, a certain amount of work is performed. By definition,

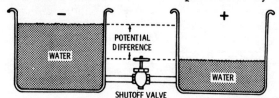

Fig. 3-1. Analogy demonstrating the principle of electron flow.

one coulomb flowing between points having a potential difference of 1 volt does the work of one joule. Then work (W) equals QE joules, where Q is the quantity of electrons given in coulombs. Power is the number of joules of work performed in t seconds. Power can be determined from the formula:

$$P = \frac{QE}{t}$$

where,
    P is the power in watts,
    Q is the quantity of electrons,
    t is the time in seconds.

Use of the fact that one coulomb moving past a point in one second is one ampere and using Ohm's law, the equation for electrical power P is:

$$\text{Power, P} = \text{IE watts.}$$
$$= E^2/R \text{ watts.}$$
$$= I^2R \text{ watts.}$$

As a matter of explanation and comparison, a power of 149,140 watts is equivalent to that of our modern automobile engine of 200 horsepower.

Resistor Circuits

The circuit of Fig. 3-2 has only one path for electron flow. Since that path has a series resistance of 6 ohms, a storage battery having an emf of 6 volts will cause one ampere of current. Therefore one

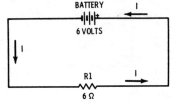

Fig. 3-2. Electrical circuit with a 6-ohm series resistance.

coulomb is passing through this resistance during each second. If this were compared with the passage of a ball through a length of pipe in 1 second, it would seem reasonable to assume that it would require 2 seconds for the ball to move through 2 lengths of pipe. Similarly, a circuit having two series resistances of 6 ohms each (Fig. 3-3) will pass 1 coulomb in 2 seconds (one-half coulomb each second). The total series resistance in Fig. 3-3 is 12 ohms. Thus, two or more resistances in series have an effective resistance equal to the sum of those resistances.

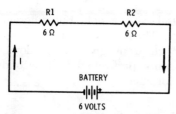

Fig. 3-3. A circuit with two 6-ohm resistances in series.

The circuit shown in Fig. 3-4 has 4 resistors in series with values of 2, 3, 4, and 9 ohms, respectively. The total series resistance is therefore 18 ohms. An emf of 18 volts acting on this circuit will cause a current of 1 ampere. Since the emf required to cause 1 ampere of current through R1 (2 ohms) is 2 volts it can be assumed that the potential difference between points A and B is 2 volts. Likewise the potential differences between points B and C, C and

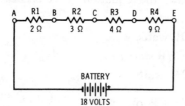

Fig. 3-4. A circuit with four series resistors of unequal value.

D, D and E, are 3, 4, and 9 volts, respectively. By adding the individual voltage drops across these resistors $(2 + 3 + 4 + 9)$ it can be seen that the sum is equal to the emf of the battery. Hence, in series circuits such as those in Figs. 3-3 and 3-4, the voltage drop across the individual resistors is divided in proportion to the individual resistance values.

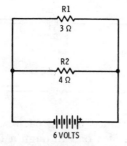

Fig. 3-5. A parallel circuit with resistors of unequal value.

Fig. 3-5 illustrates a *parallel circuit*. The parallel circuit is analogous to the multilaned highway where each line or path carries a share of the traffic.

Conductance

The ability of a material or a circuit to pass current is referred to as conductance. Earlier, conductance was defined as the reverse of resistance. Mathematically, conductance is the reciprocal of resistance.

Conductance,

$$(G) = \frac{1}{R}$$

or,

$$R = \frac{1}{G}$$

and,

$$E = \frac{I}{G}$$

The unit of conductance is the *mho* (the reverse of ohm) and the 3-ohm resistance (R1 of Fig. 3-5) presents a conductance of ⅓ mho. Similarly, a conductance of ¼ mho is presented by the path containing R2. Since the 6-volt force of the battery acts on each path, the electron flows produced are 2 and 1.5 amperes, respectively. Total current flow within the battery is then the sum of that in each path, 2 plus 1.5 or 3.5 amperes. The effective resistance of the parallel circuit in Fig. 3-5 is 1.715 ohms, which is equal to a conductance of 0.583 mho. This effective conductance (0.583 mho) is also equal to the sum of the conductances of the respective paths, or 0.333 plus 0.259 mho.

Parallel circuits can have any number of paths and those paths can present one resistance or any number of resistances in series. In the same manner, a series circuit can consist of two or more parallel combinations of resistances. Thus, the number of variations found in resistance circuits is innumerable and can be simple or highly complex.

ELECTRICAL CAPACITANCE

Just as a dead-end street blocks the flow of traffic, so does an insulator placed in series with a conductor block the flow of electrons (Fig. 3-6). Here the wire is broken, or open, and the air gap which constitutes an insulator between points A and B blocks electron flow. As the conductor from point B is connected to the negative terminal of the battery, a number of the excess electrons from this terminal move into the wire toward point B. However,

Fig. 3-6. An open circuit.

since this is a "dead end street," the electron flow toward point B is very small and lasts for only an instant. Being deficient of electrons, the positive terminal of the battery causes a similar flow of electrons from point A to the positive battery terminal.

By conforming to the basic law of electronics—unlike charges at points A and B attract each other. The excess electrons at point B tend to move toward point A where there is a deficiency. The electrons within the insulating material (air in this case) will similarly tend to move toward the point that is deficient of electrons. However, since the electrons within the insulating material are too tightly bound to the atom nucleus to be forced from their normal orbits, the external force distorts the normal orbits of these electrons (Fig. 3-7).

Dielectrics

Insulating materials having electron orbits that can be distorted by an electrostatic force act as a cushion—the greater the distortion the more elastic the cushion. When using air as the insulating material, the electrons of the nitrogen and oxygen atoms—the two principal elements of air—are tightly held within two orbits. A sulfur atom has a third orbit of electrons that is comparatively free from its positively charged nucleus. Thus, the amount of orbital

  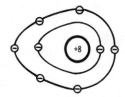

(A) Normal electron orbits when no electrostatic charge is present.
(B) Distorted electron orbits resulting from electrostatic charge.

Fig. 3-7. An electrostatic charge can affect the orbits of the electrons in a dielectric.

distortion that can occur to the electron within an air insulation is less than that of a sulfur insulation. Replacing the air insulation of Fig. 3-6 with sulfur will decrease the opposition encountered by the electrons flowing into point B, and point B can therefore acquire a greater number of excess electrons.

The insulating material separating points A and B is known as a *dielectric;* dry air is taken as a standard by which all other dielectrics are considered. The primary consideration for a dielectric is the possible degree of distortion to its electron orbits. This distortion, or elasticity, is given a numeric value known as the *dielectric constant.* As the standard dielectric, dry air is assigned a dielectric constant of 1 while that of sulfur is 2.9 and that of mica is approximately 6.

Since the electrostatic force existing between points A and B of Fig. 3-6 determines the number of electrons that will accumulate on point B, it also affects the atomic distortion. As points A and B are spaced farther apart, the effect of the electrostatic force on the distortion of the atoms will be less, and fewer electrons will accumulate at point B.

Capacitive Values

If points A and B in Fig. 3-6 are changed to plates of metal (Fig. 3-8), the "dead end street" has a greater area and hence a greater ability to acquire electrons. Plates A and B then have the capacity to acquire and repel a fixed number of electrons when

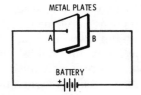

Fig. 3-8. Conductive plates connected at a break in the circuit to form a capacitor.

acted on by a given value of emf. This combination (two conductive plates and a dielectric) is referred to as a *capacitor*. By mathematical definition the capacitive value, or capacitance, of a capacitor is the ratio of the quantity of electrons acquired by its plate to the applied emf. It is expressed by the formula:

$$C = \frac{Q}{E}$$

where,

C is the capacitance in farads,
Q is quantity of charge in coulombs,
E is the electromotive force in volts.

The term farad was named for the English chemist and physicist Michael Faraday (1791-1867). A capacitor that is capable of acquiring one coulomb of charge when acted on by one volt of emf is said to have a capacitive value of one farad. More commonly used in electronics are the small units microfarad and picofarad which are one-millionth part of a farad and one-millionth part of a microfarad, respectively. (The term micromicrofarad is sometimes used instead of picofarad.)

In Fig. 3-9, resistor R (1 ohm) has been placed in series with a 0.1-microfarad capacitor across the terminals of a 100-volt battery.

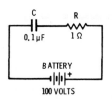

Fig. 3-9. A resistor and capacitor connected in series across a 100-volt battery.

At the start, current flow is, according to Ohm's law, 100 amperes or 100 coulombs per second. At this rate of current flow the negative plate of the capacitor will acquire a charge of 0.000002 coulomb during the first 0.02 microsecond of charging time. With this charge the capacitor has an electrostatic force of 20 volts ($E = Q/C$) which is opposing the 100-volt force of the battery and leaving an effective force of 80 volts acting on resistor R. During the next period of 0.02 microsecond, an additional charge of 0.0000016 coulomb is acquired by the capacitor; the capacitor then has an electrostatic force of 36 volts. Thus as the capacitor gains its charge, that charge reacts against further electron flow. From this it becomes apparent that there is an element of time associated with the charging and discharging rates of a capacitor. However, it is beyond the scope of this book to establish mathematically that charge-time relationship. Here it will suffice to say that the capacitor will acquire 63% of its total charge in a number of seconds equaling the product of the circuit resistance in ohms and its capacity in farads ($t = RC$). This product is known as the *time constant* of a resistance-capacitance circuit and is often used as a means of establishing the time during which an emf or a current flow may exert its effect.

Capacitor Composition

As might be expected, capacitors are found in many forms. Probably the most common is the paper tubular type, which is formed by simply rolling two strips of metal foil that are separated by a strip of paper—the paper acting as a dielectric. An excessive emf can cause this paper to "break down" and become conductive. To overcome this possibility, various forms of nonconducting and semiconducting liquids and pastes have been used as the dielectrics. Because of its high dielectric constant, mica is often used. There are also situations which require that the value of a capacitor be varied, in which case two sets of plates using air as the dielectric are constructed so that one set rotates between the other.

ELECTRICAL INDUCTANCE

Since an electron has mass (although extremely small), it can also have inertia. A moving electron will react against any attempt to alter its movement. The normal resistance of a conductor will quickly overcome the inertia of an electron moving through it; however, the inertia of an electron rotating within an orbit of the atom is opposed by little resistance and the rotation tends to continue.

Theoretically, the electron orbits of an iron atom magnetized by an external flow of electrons line up in the same plane as the electron path responsible for the magnetic condition (Fig. 3-10). Coil-

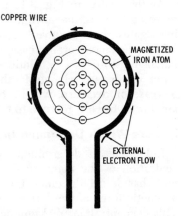

Fig. 3-10. An iron atom magnetized by an external current.

ing a wire about a piece of iron and forcing a movement of electrons through that wire will cause many of the electron orbits within the iron to align themselves in the same plane as the turns of wire (Fig. 3-11). The piece of iron is then magnetized by the electron flow—hence, an electromagnet.

Self-Induction

Should the switch in Fig. 3-11 be opened causing the current flow to stop abruptly, the iron will tend to retain its magnetic properties. Because of their inertia the magnetic-iron atoms will continue to rotate in a single plane and in a single direction. In time, however, the magnetic atoms return to their nonmagnetic state and the magnetic lines of force collapse. As this occurs these lines fold about the copper wire but concentrate only on one side. This unbalanced condition causes a momentary resumption of cur-

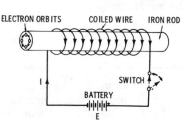

Fig. 3-11. Effect of the electron orbits of an iron rod surrounded by a current-carrying conductor.

rent flow in the wire. Since this current is created by electromagnetic induction within the physical limits of the coiled wire and its core (the atom of iron), it is termed a *self-induced* current. This flow is brought about by a self-induced emf. While the production of this force is known as self-induction, the physical ability to produce self-induction is *inductance*. Any conductor, straight or coiled, has inductance.

As the switch in Fig. 3-11 is closed, the action just described is reversed. That is, the current is opposed by those atoms of iron reacting against being converted into magnetic atoms. Therefore, the initial movement or increase in movement of current produces a balancing force (a self-induced emf) opposing such action. In other words, any change in the rate of current, whether it be an increase or decrease, will be opposed by a self-induced emf. The amount of force is relative to the change of current per unit of time.

Factors Which Determine Inductance

Inductance is determined by many factors, one being the concentration of the magnetic lines of force. For example, a straight wire has less inductance than one which is coiled, because the magnetic force of the latter is concentrated within a smaller space. This concentration of magnetic force and the corresponding inductance is further increased by adding more turns of wire to a coil. Moreover, since some materials have a greater ability than others to retain this magnetic condition, the core around which the coil is wound is also a factor in determining inductance. Hence, turns of wire about an iron core have a greater inductance than a coil of wire with a core of air. Inductance can also vary with the amount of current. That is, a large amount of current can magnetically saturate an iron core, thereby altering the inductance of the wire surrounding it.

The unit of inductance is the *henry*, acquiring its name from the American physicist Joseph Henry. The smaller units of inductance are the millihenry and the microhenry. By definition, an inductance of one henry will have a self-induced emf of one volt when the current is changed at the rate of one ampere per second. Expressing this mathematically one can define the unit in inductance by the equation:

$$e = L\frac{\Delta i}{\Delta t}$$

where,
    e is the induced voltage,
    L is the inductance in henrys,
    Δi is the change in electron flow in the time of *t* seconds.

Since the emf induced by a change in the rate of current flow through an inductance is always opposing that change, it also resists the force causing the change in the rate of current flow. Induced voltages are therefore considered negative with respect to the voltage applied to an inductance.

Fig. 3-12 shows a series circuit having a resistance of 1000 ohms and an inductance of 0.01 henry. The electron flow in this circuit is zero until the switch is closed. On closing this switch, a force

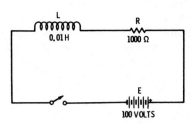

Fig. 3-12. A series resistance-inductance circuit.

of 100 volts is applied to the circuit. The resistance can initially be assumed to have no effect. After approximately 0.000001 second the inductance will develop a self-induced emf equal and opposite to the applied voltage. For an inductance of 0.01 henry to develop a self-induced voltage of −100 volts during an interval of 0.000001 second there must be a change of 0.01 ampere in current. During that first time period the current increases from zero to 0.01 ampere. However, the 1000-ohm resistance causes a 10-volt drop in the force applied to the inductance when the current is 0.01 ampere. This means that during the next interval of 0.000001 second the force applied to the inductance will be 90 volts (100 − 10), and the self-induced emf at that time will be −90 volts, the current must change from 0.01 to 0.019 ampere. In succeeding time intervals the current will be 0.01, 0.019, 0.0271, 0.0344, 0.041, 0.0469, 0.0522, etc., gradually increasing toward a maximum of 0.1 ampere. At the same time, the emf acting on the inductance approaches zero (see Fig. 3-13). The primary fact to note here is that with the

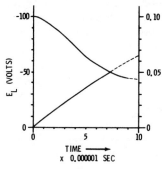

Fig. 3-13. Voltage versus current through an inductor.

presence of inductance in a circuit, there is a time lag between the application of an external emf and the maximum current.

Inductive Coupling

Current through coil L1 in Fig. 3-14 establishes a magnetic field that also surrounds the windings of L2. Any change in the rate of current in L1 alters the lines of magnetic force around L2 and induces an emf in L2. This in turn causes a current through R2. Therefore, a change in the rate of current through one inductance can, by way of a common magnetic field, produce a current within another inductance. This is known as *magnetic coupling*, or *inductive coupling*, and is the principle by which transformers operate.

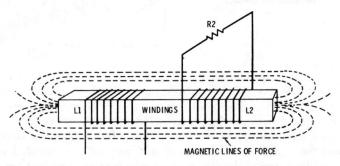

Fig. 3-14. The principle of inductive coupling between two coils.

CIRCUIT IMPEDANCE

Just as our highway speed is limited by a number of different factors, current flow is impeded by three types of restrictions.

Resistance

The time relationship of the force and the movement of a ball does coincide when rolling through a straight pipe and the force is *in phase* with the movement. The only limiting factor of the straight pipe is that of friction which does not alter the time relationship of the force and the movement. Since resistance is the electronic equivalent of friction, current through a resistance is in phase with the emf producing it. Accordingly, when an alternating emf is applied to a resistance, the maximum and the minimum values of the resulting electron flow occur at the same time as those of the force.

Inductance

In the circuit of Fig. 3-15, an emf alternating at 100 Hz and having a maximum value of 100 volts is applied to an inductance

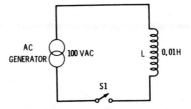

Fig. 3-15. A basic inductive circuit.

of 0.01 henry by closing switch S1. At 0.0005 second after the switch is closed, the applied voltage following a normal sine-wave variation will have dropped from 100 to 95 volts. Thus, the average emf induced by inductance during that first 0.0005 second is −97.5 volts. For an inductance of 0.01 henry to produce an induced emf of −97.5 volts, the current must change 4.875 amperes. This is according to the formula:

$$E = -L\frac{\Delta i}{\Delta t}$$

$$\Delta i = \frac{-E\Delta t}{L}$$

After 0.0005 second the current will have risen from zero to 4.875 amperes. In succeeding time periods of 0.0005 second the applied emf varying according to the trigonometric sine function has values of 81, 59, 31, 0, −31, −59, −81, −95, −100, −95, −81, etc., while the current will have values of 9.275, 12.8, 15.05, 15.83*, 15.05, 12.8, 9.275, 4.875, 0, −9.275, −12.8, etc.

The curves in Fig. 3-16 show the emf versus current occurring

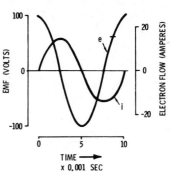

Fig. 3-16. Electromotive force versus current in the circuit of Fig. 3-15.

in the circuit of Fig. 3-15. The maximum value of the current occurs 0.0025 second after the emf reaches maximum. At a frequency of 100 Hz a period of 0.0025 second equals ¼ cycle, or 90 electrical

---

*It should be noted that the maximum current calculated by more exact methods will be 15.9 amperes.

degrees. So the current through an inductance is out of phase with, or *lags*, the emf by ¼ cycle.

Inductive Reactance

The opposition offered the flow of ac through an inductance differs from that of a physical resistance and is a reaction created by the current itself. Defined in mechanical terms, a reaction is that force produced as an opposition to an action. Electronic reaction takes the form of an emf and can be created by self-induction. Equal and opposite to the applied force, the self-induced emf has a value equaling $-L\frac{\Delta i}{\Delta t}$. From this term and known limits it is possible to conclude that the emf to current ratio (E/I) equals 6.28 fL and is referred to as the *inductive reactance*, X_L.

$$X_L = 6.28 \, fL$$

where,
    f is the frequency in hertz,
    L is the inductance in henrys.

In addition to opposing the flow of ac the inductive reactance also causes the current to lag the applied emf by ¼ cycle, or −90° (the negative sign indicating the lag). Following the normal rule governing the cause and the effect it can be assumed that the inductive reactance has an angular characteristic of +90°. This compares with the zero angular characteristic of a resistance which does not cause a change in the phase between the current and the emf (see Fig. 3-17).

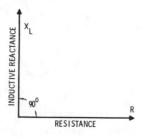

Fig. 3-17. The angular relationship between resistance and inductive reactance.

Capacitive Reactance

In Fig. 3-18, a 1-microfarad capacitor is acted on by a 100-Hz ac potential. The maximum value of the alternating emf is 100 volts. Closing switch S1 the instant the emf is zero and starting toward the positive half-cycle, the capacitor begins to charge. During the first 0.00025 second after the switch is closed the emf rises to a value of 15.6 volts and the capacitor charge (q) becomes 0.0000156 coulomb. This is equivalent to an initial current of about 0.063 ampere. But in succeeding time periods the charge within the

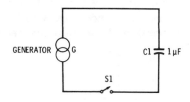

Fig. 3-18. A basic capacitive circuit.

capacitor opposes additional charge and the current decreases. When the alternating emf decreases from its maximum positive value, the current reverses direction and increases. Plotting the variations of the emf and the current (Fig. 3-19), it can be seen maximum current occurs ¼ cycle ahead of maximum emf. Therefore the current into a capacitor *leads* the emf by ¼ cycle, or 90°.

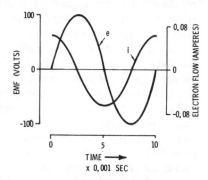

Fig. 3-19. Electromotive force versus current in the circuit of Fig. 3-18.

Like the inductor, the capacitor offers an opposition to ac. Here, the capacitive reaction is equal and opposite that of the applied emf. This opposition is termed *capacitive reactance,* and for a given capacity and frequency is expressed by the formula:

$$X_C = \frac{1}{2\pi fC}$$

where,
  X_C is the capacitive reactance in ohms,
  C is the capacity in farads,
  f is the frequency in hertz.

Reactance differs from resistance by changing the time relationship of the emf and current alternations. Capacitive reactance can be considered a negative reactance since it causes the alternations of current to lead alternations of emf.

Impedance

Opposition to alternating current, thus, may have three components—resistance, inductive reactance, and capacitive reactance. Combined, the resultant of these forms of opposition is referred to as *impedance*. Fig. 3-20 indicates in vector form the angular relationship of the inductive reactance, the capacitive reactance, and

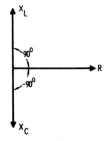

Fig. 3-20. The angular relationship among inductive reactance, capacitive reactance, and resistance.

the resistance. The inherent angular characteristics of the inductive reactance and the capacitive reactance established these two components to be contradictory forms of opposition. Ten ohms of inductive reactance in series with a capacitive reactance of six ohms has a resultant reactance of four ohms ($10 - 6$) with an angular characteristic determined by the larger value of the inductive reactance. If the capacitive reactance had been the larger, the effective reactance would have been capacitive.

The generator in Fig. 3-21 supplies an alternating current to capacitive reactance X_C in series with resistance R. If this current has a maximum value of 3.54 amperes, it will develop an alternating potential difference of 35.35 volts across the 10 ohms of capacitive reactance. But the potential difference across the capacitor will be out of phase with the current while the potential difference of the resistance will alternate in phase with the current. Table 3-1 gives the instantaneous values of the factors involved.

Plotting variations of current and those of the total potential difference (Fig. 3-22) indicates a 45° phase shift. The current alternations lead those of the emf suggesting that capacitive reactance and resistance forms an impedance with an angular characteristic of

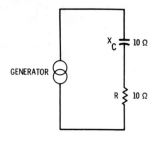

Fig. 3-21. A basic resistance-capacitance (RC) circuit.

Table 3-1. Instantaneous Values of the Factors Associated With the Circuit of Fig. 3-21

Angle	i	E_c	E_r	E_t
0	0	−35.35	0	−35.35
22.5	1.34	−32.7	13.4	−19.3
45.0	2.50	−25.0	25.0	0
67.5	3.28	−13.4	32.8	19.3
90.0	3.54	0	35.4	35.4
112.5	3.28	13.4	32.8	46.2
135.0	2.50	25.0	25.0	50.0
157.5	1.34	32.7	13.4	46.2
180.0	0	35.4	0	35.4
202.5	−1.34	32.7	−13.4	19.3
225.0	−2.50	25.0	−25.0	0
247.5	−3.28	13.4	−32.8	−19.3
270.0	−3.54	0	−35.4	−35.4
292.5	−3.28	−13.4	−32.8	−46.2
315.0	−2.50	−25.0	−25.0	−50.0
337.5	−1.34	−32.7	−13.4	−46.2
360	0	−35.4	0	−35.4

i   is the instantaneous value of current.
E_c is the potential difference developed across the capacitive reactance.
E_r is the potential difference developed across the resistance.
E_t equals E_c plus E_r.

−45°. More commonly, this angular characteristic is referred to as the *phase angle* and is dependent on the ratio of the reactance to the resistance of a circuit. When this phase angle is identified as θ:

$$\text{Tangent } \theta = \frac{X}{R}$$

According to Table 3-1 the maximum emf in Fig. 3-21 is 50 volts, and its maximum current is 3.535 amperes, meaning that impedance E/I is 14.14 ohms. The impedance is then the resultant of quadrature addition of the reactance and the resistance. *Quadrature addition* can be likened to the man who walks from one point toward

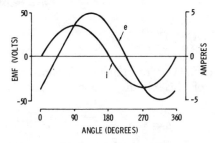

Fig. 3-22. Electromotive force versus current in the circuit of Fig. 3-21.

the north, then turns east to come to a final position northeast of the original point. The resultant of the man's walk would be as though he had taken a direct course from the original point northeast to his final position. A right-angle triangle is formed by the two components of the walking and their resultant (Fig. 3-23). The resultant or the impedance (Z) is the square root of the sum of the other two sides of the triangle.

$$Z = \sqrt{R^2 + X^2}$$

Thus, a circuit having a resistance of 10 ohms and a reactance of 10 ohms will have an impedance of 14.14 ohms.

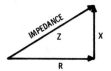

Fig. 3-23. The impedance of the circuit in Fig. 3-21 is the resultant of quadrature addition of the resistance and the reactance.

$$
\begin{aligned}
Z &= \sqrt{R^2 + X^2} \\
&= \sqrt{(10)^2 + (-10)^2} \\
&= \sqrt{100 + 100} \\
&= \sqrt{200} \\
&= 14.14 \text{ ohms}
\end{aligned}
$$

Fig. 3-24 shows a circuit having a resistance of 4 ohms, an inductive reactance of 4 ohms and a capacitive reactance of 1 ohm. Such

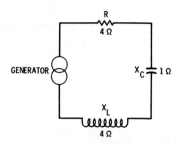

Fig. 3-24. A basic circuit consisting of inductance, capacitance, and resistance.

a circuit has an effective reactance of 3 ohms which is inductive. The impedance of this circuit is 5 ohms, and is determined as follows:

$$
\begin{aligned}
Z &= \sqrt{R^2 + (X_L - X_C)^2} \\
&= \sqrt{16 + 9} \\
&= 5 \text{ ohms}
\end{aligned}
$$

If the frequency of the applied emf in Fig. 3-24 is changed to half of that previously assumed, the 2 reactances (capacitive and inductive) will be equal to 2 ohms. Since the 2 reactances at this frequency are equal and opposite, they cancel each other and the resulting impedance is the value of the resistance (4 ohms).

Resonant Circuits

Fig. 3-25 shows a series circuit having a resistance of 1 ohm, a capacitive reactance of 4 ohms, and an inductive reactance of 1 ohm. With a current of 1.0 ampere the voltage developed across resistance R is 1.0 volt, that voltage across capacitive reactance

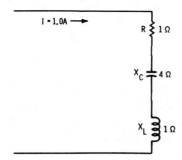

Fig. 3-25. Series resonant circuit.

X_C is 4.0 volts, and that across inductive reactance X_L is 1.0 volt. But as shown by the vectors of Fig. 3-26, the voltage across the inductive reactance E_L leads that across the resistance E_R by 90° while that voltage across X_C lags by 90°. Upon adding these three voltages vectorally we find the result to be an effective voltage E of 3.16, that lags that across R by less than 90°. This gives an impedance Z equivalent to E/I, or 3.16/1, or 3.16 ohms. If the frequency is doubled, the inductive reactance and the capacitive reactance become equal. Under these conditions the effective voltage E is only that across resistance R, or 1.0 volt, and the

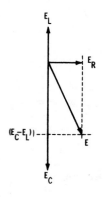

Fig. 3-26. Vector representation of voltage in a series resonant circuit.

impedance Z is that of the resistance R, or 1 ohm. Plotting values of impedance against the various frequencies (Fig. 3-27) indicates that at one frequency—the *resonant frequency*—the impedance decreases to a small value. Referred to as a *series resonant circuit*, the circuit shown by Fig. 3-25 is selective of the frequency passed.

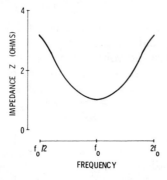

Fig. 3-27. Series resonance frequency response.

Fig. 3-28 shows a *parallel resonant circuit*. Applying a constant voltage of 10 volts produces a current of 1.0 ampere through the capacitive reactance of 10 ohms, 4 amperes through the 2.5-ohm inductive reactance, and 1 ampere through the 10 ohms of resistance. Following much the same mathematics as with the series resonant circuit, the impedance variations become evident as the frequency is altered. Fig. 3-29 shows that the parallel resonant

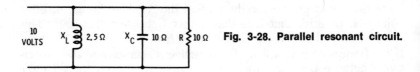

Fig. 3-28. Parallel resonant circuit.

impedance increases to a high value at its resonant frequency. Theoretically, the impedance of a parallel resonant circuit without the resistance R has an infinite impedance at the resonant frequency. However, the inductance always has enough resistance to keep the resonant impedance from being infinite. In a sense, the parallel resonant circuit acts as a tank that traps the alternating current and accepts only a given amount of the alternating current. Often referred to as a "tank circuit" the parallel resonant circuit becomes "filled" and then "spills over" into whatever path is provided. Consequently, the tank circuit is selective of the frequency it rejects or opposes.

Resonance, in either a series or parallel combination, occurs at only one frequency for any given combination of capacitance and

46

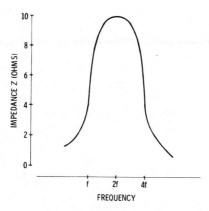

Fig. 3-29. Parallel resonance frequency response.

inductance. That resonant frequency may be determined by the formula:

$$f_o = \frac{1}{6.28\sqrt{LC}}$$

where,

    f_o is the resonant frequency,
    L is the inductance in henrys,
    C is the capacitance in farads.

REVIEW QUESTIONS

1. Explain why the atomic structure of copper allows a free electron to move rapidly compared with movement in carbon.
2. What is the name of the characteristics of materials that is the opposite of resistance?
3. What is the unit of measure for electromotive force? *Volts*
4. If one coulomb, 6,280,000,000,000,000,000 electrons, is moving past a given point each second, what is the rate of current flow given in common units? *AMPS.*
5. What is potential difference? *A Access of elec. — +*
6. How much power is being consumed by a circuit that passes 10 amperes of current when acted on by a force of 6 volts?
7. If a series circuit has 2 resistances, one of 5000 ohms and one of 9500 ohms, what is the total effective resistance? *14500*
8. A parallel circuit has 1 branch or path with 12 ohms of resistance and another path with 4 ohms resistance. What is the effective resistance? *19 ohms*
9. Why are the electron orbits between two charged points distorted? *Prevented from Moving freely*

10. What is a dielectric? *an insulating material with known Distortion*
11. What is the dielectric constant of dry air? Of sulfur? Of mica? *2.9* *6*
12. What is a capacitor?
13. What is inertia?
14. Will the inertia of an electron within an orbit of an atom affect its movements?
15. Describe what happens to the atoms of iron when surrounded by a coiled wire with electrons flowing through it.
16. What is self-induced electromotive force?
17. What is inductance?
18. Name some of the factors that affect the inductance of a given coil and core.
19. What is the unit of inductance? *Henry*
20. Why is the electromagnetically induced voltage considered negative with respect to the applied voltage?
21. Does the maximum current in a circuit having inductance occur at the same instant the external emf is applied?
22. Can a change in the rate of current through one inductance produce a current within another inductance? Explain.
23. What is meant by a force being out of phase with the movement?
24. Is the alternating current through a resistance in phase or out of phase with its emf?
25. Does the alternating current lead or lag the alternations of the emf when moving through an inductance?
26. How much out of phase is the current through an inductance acted on by an alternating emf?
27. Define inductive reactance. State the formula for inductive reactance.
28. Define capacitive reactance. State the formula for capacitive reactance.
29. What is impedance? Name its three components.
30. What is series resonance? How many resonant frequencies are there for a given combination of capacity and inductance?
31. Describe the tank circuit. Is the impedance of a tank circuit high or low at its resonant frequency?

CHAPTER 4

Electron Tubes

Electron tubes have, to a great extent, been replaced by transistors and other solid-state devices. However, the fundamental principles, usage and historical significance of electron tubes still warrant some notation. Historically, the electron tube—the radio tube, the vacuum tube, the valve, etc.—was the originating point of electronics. Its operating principles vividly illustrate many of the basics of electrostatic charges and their effects. In many of today's homes, the only electron tubes used presently are the picture or cathode-ray tube of the television receiver and the magnetron of the microwave oven. However, the radio and television stations are still using a great number of electron tubes to produce their radiating power (solid-state devices are not capable of dissipating the great amount of heat involved).

THE DIODE

At some period during the experimenting which led to the discovery of the incandescent lamp, Thomas A. Edison placed a metal plate inside an evacuated glass bulb or tube. He noted that when a wire was connected between this metal plate and the positive side of the battery furnishing power to the filament, current occurred within the wire. But no current developed when the plate was connected to the negative battery terminal. Now known as the "Edison effect," it provided the basis for the entire electron theory.

It was found that as the filament wire became heated, a number of electrons were emitted from it into the vacuum of the glass envelope. A metal plate in that same vacuum and having a déficiency of electrons (positively charged) would attract those electrons emitted from the heated filament. If the metal plate were charged with a surplus of electrons (a negative charge) it would repel any electrons coming from the filament.

Referred to as a *diode*, the two elements within the vacuum form a semiconductor—passing current freely in one direction while opposing it in the other direction. The diode also illustrates vividly the laws of attraction and repulsion. Fig. 4-1 shows the diode tube and the relative positions of the filament or cathode, from which electrons are emitted, and the anode or plate. Diode tubes find use primarily in rectification—converting alternating current to a pulsating form of direct current. Such rectification of higher frequency currents used in broadcasting removes the intelligence.

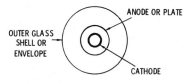

Fig. 4-1. Relationship between the elements of a diode tube.

Marchese Guglielmo Marconi, Italian inventor (1874-1937), patented the first wireless telegraph on June 2, 1896. The signal of the wireless telegraph consisted of an interrupted electron flow that alternated at a fairly high frequency. Since the telegraph sounder (the electromagnetic mechanical device shown in Fig. 4-2) could not follow those rapid alternations, a rectifier was used to eliminate half of the cycle. The telegraph sounder then reacted to the average value of the resulting pulsating electron flow. Originally a crystal of galena or quartz having semiconductor characteristics was used to rectify the wireless telegraph signal. This converted the high frequency current to a pulsating form of direct current causing the telegraph sounder to function.

In 1897, the English electrical engineer J. A. Fleming improved wireless telegraphy by replacing the semiconductor crystal with a

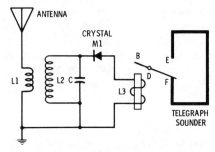

Fig. 4-2. A telegraph sounder and associated circuitry.

diode electron tube. The detector was then referred to as the *diode detector*. Another improvement was made by replacing the telegraph sounder with the telephone receiver or headphone (Fig. 4-3).

THE TRIODE

In 1906 an American inventor named Lee de Forest introduced a tube having the ability to amplify an ac signal applied to it. He referred to this as the *audion* tube. By connecting this tube into a circuit supplied with the proper voltages, it could amplify weak telephone signals sufficiently to operate an earphone or even a speaker. The audion was a three-element tube, and is now referred to as the *triode tube*.

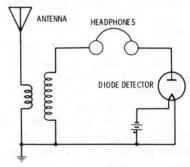

Fig. 4-3. Basic receiver circuit using a diode detector and headphones.

The third element of the triode, called a grid, is a screen of fine wire placed between the cathode and the anode (Fig. 4-4). Both the anode and the grid are electrostatically charged to produce fields of force acting on those electrons emitted from the cathode. With the grid placed closer to the cathode, its electrostatic effect upon electrons coming from the cathode is greater. Thus, a comparatively small negative charge on the grid overcomes the electrostatic force pulling electrons toward the anode. A small negative grid charge reduces the anode current considerably. If the grid charge is varied by a generator G, as in Fig. 4-5, larger but similar variations occur across the load resistance R_L. In effect, the changes in the grid charges are increased, or *amplified,* in the anode circuit. An alternating emf having a maximum value of 1.0 volt from generator G may thus become a 20-volt variation across R_L. If the triode tube is

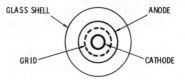

Fig. 4-4. Relationship among the elements of a triode tube.

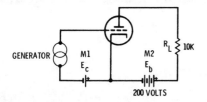

Fig. 4-5. Basic electron-tube amplifier.

arranged to give large variations of current, rather than voltage, the circuit of Fig. 4-5 becomes a power amplifier—load resistance R_L being replaced by a transformer-speaker combination, or other components, dissipating the power in some means other than heat.

The four-element tetrode, the five-element pentode and the seven-element heptode used additional grids and screens to reinforce the action of the other elements. With its seven elements, the heptode became a combination rf amplifier, oscillator and converter. In modern tubes, it is common to find more than one diode, triode, tetrode, pentode, heptode or combinations within a single vacuum enclosure. Actually we are describing only the *basic* types of electron tubes; there are too many to name.

Some of the tubes used for power amplifiers in eras past were as much as five feet in length and used circulating distilled water to dissipate heat. However, these were still of the triode type and many of the more modern transmitter tubes are also triodes. Uhf television transmitters, using magnetrons, are probably the primary exception to the triode usage.

CHAPTER 5

Solid-State Physics

Much of the study of solid-state physics began during World War II. In those war years germanium crystals were used for the detection and rectification of the high-frequency signals associated with radar. Later, in 1948, three scientists employed with Bell Telephone Laboratories—Dr. Walter H. Brattain, Dr. John Bardeen, and Dr. William Shockley—made a detailed study of these crystals and their functioning. For these discoveries they received the Nobel prize in physics.

Those discoveries dealt with the electrical properties of crystalline materials and their atomic structures. Along with the electronic discoveries, methods were also found by which germanium and silicon could be supplied in a nearly pure state. From these principles the various solid-state semiconductors and transistors have developed.

CRYSTAL STRUCTURE

While crystalline materials such as germanium, silicon, and quartz are basically composed of molecules and atoms like other materials, the crystal has a fixed pattern of construction. Each atom of silicon or germanium has four electrons within its outer or valence shell, as shown by Fig. 5-1. A basic rule of chemistry is that the atom tends to fill its valence shell with eight electrons by sharing valence electrons of other atoms. Thus, each silicon or germanium atom will share one of its valence electrons with each of four other atoms. In turn each of the four other atoms shares one of its valence electrons with the first atom. Fig. 5-2 shows that

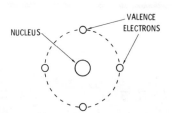

this forms a very regular latticelike pattern. In contrast, an atom of iron has three electrons in its valence shell and its molecular pattern is not as regular as that shown by Fig. 5-2. Iron is a conductor of electricity, or electron flow, while the silicon or germanium crystal is rather indifferent as to whether or not it will conduct electricity.

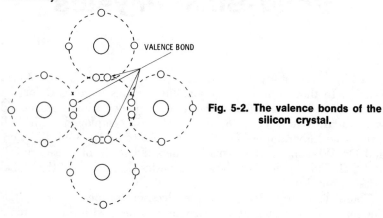

Fig. 5-2. The valence bonds of the silicon crystal.

DOPED CRYSTALS

Oddly enough one reason why it was essential to produce germanium and silicon in a pure state was so that it could be made impure. The silicon or germanium found in the natural state has so many different impurities that addition of other impurities gave very uncertain results. Silicon or germanium in its pure state has the molecular structure shown in Fig. 5-2. If an atom of antimony or arsenic having five electrons in its valence shell is injected into a silicon crystal, four of those valence electrons of the antimony are shared with four of its neighbor silicon atoms. However, the fifth valence electron of the arsenic is then detached and becomes a free electron. Such a crystal with free electrons is referred to as a negative or n-type crystal, and is shown by Fig. 5-3.

When the crystal is "doped" with aluminum or gallium atoms, a positive or p-type crystal is developed. The aluminum atom has

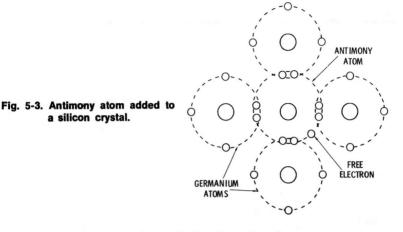

Fig. 5-3. Antimony atom added to a silicon crystal.

ANTIMONY ATOM

FREE ELECTRON

GERMANIUM ATOMS

three electrons in its valence shell. Thus, the aluminum atom can share one electron with each of three silicon atoms. One silicon atom, as well as the aluminum atom, is still lacking a "desired" electron within its valence shell. This space that remains unfilled is referred to as a *hole*. A hole can only be filled by an electron and therefore has a characteristic opposite the electron. In other words, the hole is a positively charged element. When an electron does fill a hole, the space from which that electron escaped becomes a hole. Therefore the hole effectively moves. A conductor, such as copper, has too many free electrons to sustain a hole, and the hole cannot exist within an insulator because of its molecular structure. So, the hole exists only within the confines of doped crystal as shown in Fig. 5-4. The crystal having these positively charged holes is the positive or p-type crystal. The majority current carriers in p-type materials are holes, and the minority carriers are electrons.

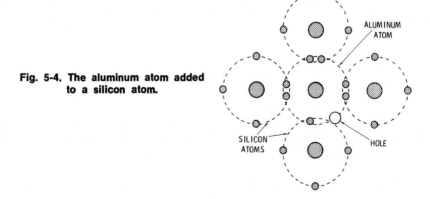

Fig. 5-4. The aluminum atom added to a silicon atom.

ALUMINUM ATOM

SILICON ATOMS

HOLE

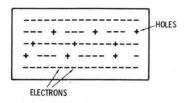

Fig. 5-5. N-type crystal with its few holes and abundance of electrons.

In actual crystals both free electrons and holes are present. But, the n-type crystal has a great many more free electrons than holes, and the holes in a p-type crystal far exceed its free electrons. Fig. 5-5 shows the few holes ($+$) among the abundance of electrons ($-$) within an n-type crystal. The drawing of the p-type crystal, Fig. 5-6, indicates the relative number of holes and electrons. The majority carriers in n-type materials are electrons, and the minority carriers are holes.

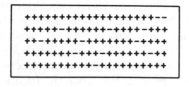

Fig. 5-6. P-type crystal with its few electrons and abundance of holes.

Diffused Crystal Material

Since crystals are literally "grown" somewhat like the stalk of corn coming from the soil, it is possible to add varying amounts of impurity. Thus, a crystal may have different concentrations of free electrons or holes at the different levels of its structure. The various levels of concentration may be sharply defined so that one section of the crystal has a high concentration of holes while an adjacent section has very few holes. Or, the concentration of holes may be *diffused* or graded from a high concentration of many holes per cubic centimeter to a low concentration at the opposite side of the crystal. Fig. 5-7 illustrates the diffused concentration of holes within a diffused crystal.

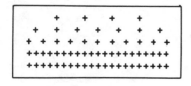

Fig. 5-7. The diffused concentration of holes within a diffused crystal.

Resistance Variation

Since the resistivity of a material—its inherent resistance per unit of volume—is directly proportional to its number of free electrons or holes, the concentration of these current carriers within a crystal also establishes its resistivity. The pure germanium or silicon crystal, with few current carriers, has a fairly high resistivity while the doped crystals of Fig. 5-5 and Fig. 5-6 have lower resistivity. As the concentration of impurity, and current carriers, can be varied, the resistivity can also be concentrated. Diffused crystal material provides a means of establishing a desired current path. That is, in the crystal of Fig. 5-7 current will concentrate along the lower portion where the current carriers are more numerous. Such variation in resistivity also provides a means of constructing solid-state resistors.

SEMICONDUCTOR ACTION

A semiconductor, by electrical definition, is any device or substance that conducts electrical current readily in one direction while opposing current flow in the opposite direction. Fig. 5-8 shows a block of pure germanium crystal on a metal plate, which is connected to the positive terminal of a battery. With the negative terminal of the battery connected to a point contact on the side of the crystal opposite the plate, meter M will indicate a passage of current. With the connections made as shown in Fig. 5-8, the large surface of the metal plate can readily attract all of those electrons entering the crystal by way of the point contact. With the connections reversed as in Fig. 5-9, the positive terminal connected to the point contact attracts very few of the electrons toward the smaller

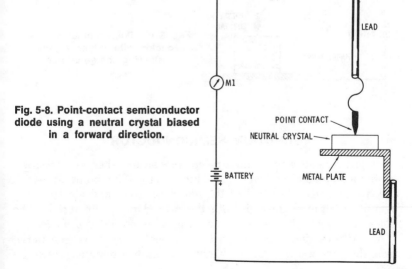

Fig. 5-8. Point-contact semiconductor diode using a neutral crystal biased in a forward direction.

57

area of the point contact, and meter M will show little or no current. Thus as connected in Fig. 5-8 the point contact, crystal, and plate combination freely passes current, while in the circuit of Fig. 5-9 the combination opposes current passage—a semiconductor.

If the pure germanium block of Fig. 5-9 is replaced with a p-type crystal, the holes will be readily attracted to the negative metal plate where an exchange of holes to electrons occurs. But, the point contact cannot attract all of the electrons entering the crystal at the plate. In fact the holes literally smother those electrons and

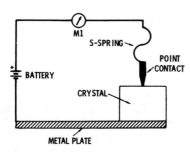

Fig. 5-9. Point-contact semiconductor diode using a neutral crystal biased in a reversed direction.

hold them close to the surface of the crystal lying against the plate. This builds up a negative charge or barrier just inside the crystal's surface that opposes and stops additional electron flow from the battery. Fig. 5-10 shows the charge barrier developed inside the crystal's surface.

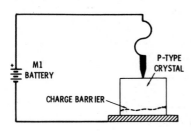

Fig. 5-10. Point-contact semiconductor with p-type crystal, showing charge barrier.

JUNCTION SEMICONDUCTOR

By growing an n-type and a p-type crystal together, another form of semiconductor is developed (Fig. 5-11). The point at which the two crystals join is called the *junction*. Holes and electrons are evenly distributed through both the n- and p-type materials. The holes and electrons are in a random movement called *diffusion*.

Majority carrier electrons (free electrons) in the n-type material near the junction are just as likely to cross the boundary into the

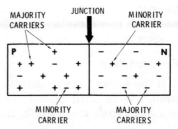

Fig. 5-11. Junction semiconductor showing initial conditions.

p region as to move further into the n region. When this starts to happen, these migrating electrons and holes will recombine close to the junction. The electrons that have entered into combination with holes have left behind positively charged atoms. This results in a situation as depicted in Fig. 5-12. The plus signs in the n region near the junction are the positively charged atoms which lost their free electron. The minus signs in the p region are the negatively charged atoms which had their holes filled with electrons. This region is known as the *depletion* region, because there is a depletion, or lack of holes and electrons in this area. The important thing to note is that any additional electrons that would diffuse from the n region to the p region are repelled by the charged atoms. Also,

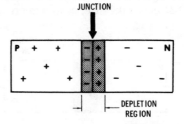

Fig. 5-12. Results of migration and recombination.

holes attempting to move from the p region to the n region are repelled by the positively charged atoms. For this reason, the electric field created by the charged atoms (called ions) in the depletion region is called a barrier.

Reverse Bias

When an external battery is connected to a pn junction, as shown in Fig. 5-13, with the positive terminal connected to the n region and the negative terminal to the p region, the junction is said to be reverse biased. Holes in the p region are attracted to the negative battery terminal away from the junction. Electrons in the n region are attracted to the positive terminal of the battery away from the junction. Thus, both types of majority carriers move away from the barrier, leaving behind more charged atoms to add to the junction barrier. This continues until the barrier charge equals the potential of the external battery, then the current stops. This condition is conventionally called reverse bias or back bias because it offers maximum resistance to an external flow of majority carriers.

Forward Bias

If the battery connections in Fig. 5-13 are reversed, the junction would be forward biased. This is shown in Fig. 5-14. The plus terminal repels the holes in the p region toward the junction, and

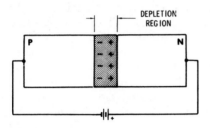

Fig. 5-13. Reverse-biased pn junction.

the negative terminal repels the electrons in the n region toward the junction. Some of these holes and electrons enter into the depletion region and combine. When this happens, an electron enters the n-type material from the external wire under the influence of the battery. Likewise, a hole is created in the p-type material by an

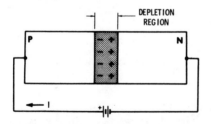

Fig. 5-14. Forward-biased pn junction.

electron breaking a co-valent bond and entering the positive terminal of the battery. Under the influence of the external battery, electrons flow in in the external circuit. This condition is known as forward-biasing the junction. A pn junction that is forward biased conducts current; a junction that is reverse biased does not.

REVIEW QUESTIONS

1. For what company did the pioneers in solid-state physics work? In addition to the electronics discoveries about crystal, what else was learned by these pioneers?

2. How does the molecular structure of a crystal differ from that of other materials?

3. If a crystal has a number of free electrons, what type crystal is it? What is a hole? What is a p-type crystal?
4. If a crystal has both free electrons and holes but the number of holes exceeds the number of free electrons, what is the crystal type?
5. How does diffused crystal material differ from ordinarily doped crystal?
6. What is a semiconductor? Give the physical description of a junction semiconductor.
7. If the p-type section of a junction diode is connected to the positive terminal of a battery, will it pass current freely? What is the reverse-bias condition?
8. What is the depletion region?

CHAPTER 6

The Solid-State Diode

Electronics began with wireless telegraph using the galena crystal with a "cat-whisker" point contact as a detector. We advanced to the use of vacuum tube diodes for detecting the wireless. Then during World War II with the advent of the ultrahigh frequencies it became necessary to again use crystal-type detectors in communications equipment.

BASIC TYPES

Following the "cat-whisker" principle an early version of the solid-state diode used a block of germanium crystal mounted on a metal plate with the sharpened point of an S-shaped spring making contact on the opposite side, as shown by Fig. 5-8. Stated as simply as possible the point contact diode functioned because the large surface of the metal plate could attract electrons from the point contact and through the crystal. In the opposite direction the point contact with its smaller area could not attract electrons through the crystal. And while this combination presented very little capacitance to affect its use as a detector in ultrahigh frequency equipment, its inherent mechanical instability was definitely a great disadvantage.

Use of a doped crystal in the point-contact version of the diode improved the efficiency, but still greater efficiency and mechanical perfection has come from the *junction-type diode*. The junction-type diode consists of a section of n-type crystal making a junction with a section of p-type crystal, as was explained previously in Chapter 5.

SILICON RECTIFIERS

Silicon rectifiers are junction-type diodes. With little or no capacitance and a 99% rectification efficiency the silicon rectifier is ideal for the rectification or detection of modulated signals. But the silicon rectifier also can have electrical ratings permitting it to carry hundreds of amperes with voltage levels greater than 1000 volts. Silicon rectifiers have excellent life characteristics as well as being small and lightweight.

ZENER DIODES

At a specific reverse voltage applied to a diode, known as the breakdown or zener voltage, a very sharp increase in the reverse current occurs. In many applications, rectifiers can operate at this zener voltage. However, operation beyond this reverse breakdown level can produce runaway current and permanent damage to the crystal. A zener diode might be used in any circuit where it is desirable to prevent current flow until the voltage has reached a given level. The voltage at which the reverse breakdown level occurs can be established by varying the concentration of impurities within the crystal sections.

SILICON CONTROLLED RECTIFIERS

The silicon controlled rectifier is one of a group of diodes known as thyristors that can be switched between ON and OFF conducting states. The silicon controlled rectifier (SCR) has a four-layer pnpn construction as shown by Fig. 6-1A. With the anode p-section made negative, the natural charge barriers are intensified and the unit is reverse biased. However, a positive voltage applied to the *gate* p-section (Fig. 6-1B) breaks down these charge barriers to start a reverse current limited only by the external circuit. After the SCR is triggered by this signal to the gate, the current from the

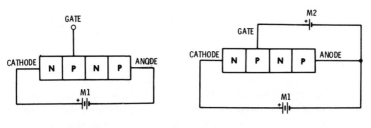

(A) Reverse biased. (B) Gate voltage applied.

Fig. 6-1. Construction and basic circuit of silicon controlled rectifier.

anode to the cathode continues until stopped by some outside action. Thus, the SCR may be used as a switching device.

TUNNEL DIODES

Tunnel diodes consist basically of two very highly doped crystal sections. This high impurity density makes the charge barrier very narrow and the electrons and holes are able to literally "tunnel" under the charge barrier. Accordingly the tunnel diode conducts in either the forward or reverse bias direction with the application of very small voltages. However, with forward biasing the current increases with the increase in voltage until peak current is realized. Further increase in the voltage reduces the tunneling and the current to give an effect of a negative resistance. Still further increase in voltage brings the device into a condition of operation like other semiconductors. Fig. 6-2 graphically shows the various current variations occurring with different voltage levels. The negative-resistance area of Fig. 6-2 is of particular interest for use in amplifier and oscillator circuits, but is beyond the limits of discussion here.

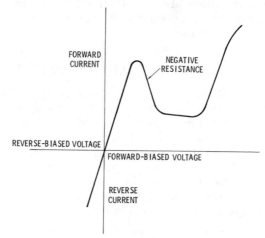

Fig. 6-2. Operating characteristics of the tunnel diode.

TUNNEL RECTIFIERS

Tunnel diodes also function as rectifiers, and due to the tunneling effect can provide rectification at much smaller voltages. However, as shown by the graph of Fig. 6-2 this small voltage rectification occurs when the tunnel rectifier is biased in the reverse direction. Thus, tunnel rectifiers are sometimes called "back diodes." Due to this high speed capability and superior rectification characteristics, the tunnel rectifier can be used to provide coupling in one direction, while isolating in the opposite direction.

VARACTOR

A varactor is a pn junction semiconductor diode designed for low losses at high frequencies. Its capacitance varies with the applied voltage. In the normal semiconductor diode, efforts are made to minimize inherent capacitance, while in the varactor, this capacitance is emphasized. Since the capacitance varies with the applied voltage, it is possible to amplify, multiply, and switch with this device.

PHOTOCONDUCTIVE CELLS

Somewhat different from the solid-state diodes described above, the *photoconductive cell* consists of cadmium sulfide or cadmium-sulfo-selenide deposited in a snakelike fashion upon a ceramic base and mounted in a glass covered case. The crystalline cadmium sulfide is a semiconductor. But as the intensity of the light striking this crystalline material increases, its resistance decreases. Typical ratings show a photoconductive cell to have a resistance of 10 megohms in total darkness and less than 300 ohms at 92 footcandles. One use of this photoconductive cell is the establishment of base and emitter currents through a transistor and, in turn, through a solenoid. This solenoid magnetically acts upon armatures to open or close the aperture of a camera to assure proper film exposure. When the light falls below a specified level a switch is also closed to the flash bulb, by this photoconductive cell arrangement.

LIGHT-EMITTING DIODES

In our television receivers a beam of electrons striking the picture tube screen produces a radiation that we detect as light. In many solid-state devices there is a similar radiation produced as the electrons and the holes cross a pn junction. With the addition of various crystal salts these radiations become detectable as light. While the usage of these light-emitting diodes seem innumerable, a primary use is within luminescent displays of letters and numerals.

PHOTOVOLTAIC CELLS

In a manner opposite that of the light-emitting diode, the photovoltaic cell converts light into electrical energy. A photovoltaic cell is basically a very thin (0.002 to 0.006 inch) wafer, or strip, of selenium or silicon. Although the nature of light is not fully understood, it is known to be a form of radiating energy with very short wavelengths. Such waves entering the confines of a selenium wafer cause its electrons to move toward the surface on which the light wave entered. A thin silvered strip on this front surface acts as a collector and conductor of those electrons. The opposite or back surface of this selenium wafer is completely silvered to form its positive terminal. Because of the difference in molecular structure the electrons of a silicon wafer move toward its back surface when the exposed front surface is illuminated. Thus, the selenium or silicon wafer combinations are capable of supplying electrical energy when exposed to light, much as a generator or battery. The space craft used by our astronauts use many of the photovoltaic cells to supply many of their needs for electrical energy. And the communication satellites, by which television is beamed around the earth, are also powered by these photovoltaic or solar cells. Closer to home is the flash camera using photovoltaic cells instead of the usual "flashlight"-type battery.

Regrettably the space allotted has not allowed for a full discussion of the various solid-state diodes. Many of these diodes are far beyond the scope of this book, but it is hoped that a reasonable introduction has been given.

REVIEW QUESTIONS

1. Describe the point contact diode; the junction-type diode.
2. Do silicon rectifiers have good rectification efficiency? What is their life expectancy?
3. Give another name for the reverse breakdown voltage of silicon rectifier.
4. What is a silicon controlled rectifier? How many crystal sections does it have? What is the gate?
5. What is particularly interesting about the tunnel diode?
6. Why are tunnel rectifiers sometimes called "back diodes"?
7. Is the varactor equivalent to a capacitance or to an inductance in series with a resistance?
8. Does the resistance of a photoconductive cell increase or decrease when the light intensity increases?
9. Name one use for light-emitting diodes.

CHAPTER 7

Transistors

Development of the transistor began in 1949 with the studies of Brattain, Bardeen, and Shockley of Bell Telephone Laboratories. Today we have many types of transistors but their primary classifications may be listed as *bipolar* and *unipolar*. Bipolar transistors depend upon the interaction of both the positive type and the negative type-charge carriers—holes and electrons. In contrast the unipolar transistor depends upon the action of only one type charge carrier.

BIPOLAR TRANSISTORS

The bipolar transistor is a construction of three crystal sections, as shown in Fig. 7-1. One of these crystal sections is very thin and is referred to as the *base*. The base crystal section is of one type of crystal while the thicker sections on the other side are of the opposite type. So if the base is made of p-type crystal, the sections on either side will be n-type crystals. One of these thicker crystal sections is known as the *emitter*, while the other is the *collector*. Since battery polarities must be established with regard to the crystal type arrangement, the transistor is also identified by that arrangement. For example, the transistor shown by Fig. 7-1 is an npn transistor while a pnp transistor is shown by Fig. 7-2.

For the sake of explanation, Fig. 7-3 shows a transistor having a very thick base section. Such a transistor is, in effect, two semiconductor diodes with their associated junctions and charge barriers. As shown in Fig. 7-3 the p-type emitter section is made positive with respect to the n-type base section. This is the forward

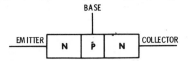

BASE

EMITTER | N | P̂ | N | COLLECTOR

Fig. 7-1. Basic npn transistor construction.

biased condition for the emitter-base junction and both holes and electrons cross that junction. In contrast, the base-collector junction is reverse biased by battery M2 and does not pass current. However,

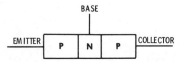

BASE

EMITTER | P | N | P | COLLECTOR

Fig. 7-2. Basic pnp transistor construction.

there are many more holes moving from the emitter into the base section than are being filled by the electrons coming from the negative terminal of battery M1. And if the base is thin, as shown in Fig. 7-4, many of the holes from the emitter will break through the base-collector junction. After breaking through this base-collec-

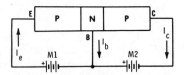

Fig. 7-3. A pnp transistor with a thick base section.

tor junction the holes will move to the collector terminal to be filled by electrons from the negative terminal of battery M2. In effect, the collector current I_C originates at the emitter terminal and passes through the emitter, base, and collector crystal sections, through batteries M2 and M1, and back to the emitter terminal.

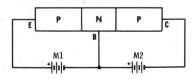

Fig. 7-4. Normal pnp transistor operation.

Besides the collector current there is also the base current I_B and the emitter current I_E. With the three currents moving in or out of the transistor, its function is that of an electrical junction. And just as the number of automobiles moving into an intersection must equal the number of automobiles moving out of that intersection, the current going into an electrical junction must equal that going out. Accordingly, the net effect of all currents in and out of

68

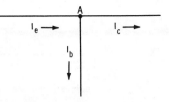

Fig. 7-5. Junction equivalent of transistor.

an electrical junction, such as that at (A) in Fig. 7-5, must equal zero.

$$I_C + I_E + I_B = 0$$

If we assume that the emitter current I_E flows out of junction (A), it can be taken to be negative. By the same reasoning the collector current I_C and the base current I_B are positive.

$$I_C + (-I_E) + I_B = 0$$
$$I_E = I_C + I_B$$

That is, the current flowing into the emitter I_E equals the sum of the currents flowing out of the base I_B and out of the collector I_C. With an npn transistor the currents are reversed in direction but the above equation is still true.

The junction between the base and the collector has a much greater surface than the wire lead attached to the base and a great many more electrons or holes can be attracted to the junction. In a typical transistor the ratio of holes, or electrons, attracted to the base-emitter junction to those attracted to the base lead might be 100 to 1. That is, the collector current may be 1.0 mA while the base current is only 0.01 mA. If the base current is increased to 0.011 mA, the collector current may be expected to increase to 1.1 mA (assuming that the I_C/I_B ratio remains constant. In this case the I_C/I_B ratio is equal to 100. By the same reasoning, the collector current to emitter current ratio, I_C/I_E, is equal to the ratio of collector current to the sum of the base and collector currents.

$$I_C/I_E = I_C/(I_B + I_C)$$

Structures

Transistor construction is primarily the formation of two parallel pn junctions with proper spacing and controlled impurity levels. Pn junctions were formed in the early point contact transistors by electrical pulsing. Grown junction transistors involve the pulling of a crystal from a molten source that is changed to differing levels of impurity. Such growing produces a large crystal having the necessary pnp or npn sections that are then sliced into a number of small-area transistors. Fig. 7-6 shows the basic structure of a grown-junction transistor.

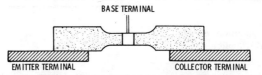

Fig. 7-6. Basic structure of a grown-junction transistor.

Fig. 7-7 illustrates the alloy-junction type of transistor. This consists of dots of impurities placed upon opposite sides of a base wafer of n-type or p-type material. After proper heating, the dots of impurities become alloyed into the base wafer to form regions for the emitter and the collector.

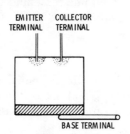

Fig. 7-7. Structure of alloy-junction transistor.

By adding a second layer to the base wafer, the drift-field transistor of Fig. 7-8 is created. Actually the heating fuses these two base layers together until the completed base region has a graded impurity level. Such a graded or varying impurity level produces a built-in voltage that speeds current flow as well as reducing the capacitive effect. This increases the frequency range of operation for the transistor.

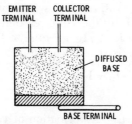

Fig. 7-8. Drift-field transistor construction.

The latest advance in transistor structure is the *diffused-junction transistor* or the *epitaxial transistor*. Diffused-junction transistors are also referred to as "homotaxial" structures. Homotaxial refers to a precise system of producing alloy-junction transistors, while the epitaxial is more of a grown-junction precision device. Much of this construction involves photolithographic and masking techniques. In simpler terms, the process is that of masking off certain areas and using acid to etch away portions of the crystals. Then other layers

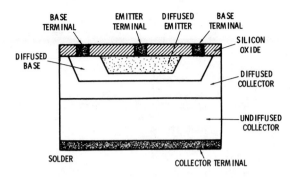

Fig. 7-9. Triple-diffused planar-type transistor.

are put on and etched away in succession to form varied configurations. Fig. 7-9 shows the cross-section of a triple-diffused planar type transistor, while an epitaxial-base type is shown by Fig. 7-10. The final package with proper heat sinks can deliver hundreds of watts of power and currents of tens of amperes.

Basic Circuits

All circuits using transistors are of three basic forms—common-base, common-emitter, and common-collector. In general when a small signal is applied to the input of a transistor circuit an amplified

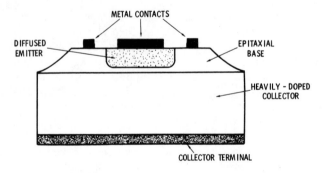

Fig. 7-10. Epitaxial-base type transistor.

reproduction of that signal appears across the output of that circuit. Fig. 7-11 shows the common-base circuit having the signal introduced by generator G within the emitter-base loop. As shown by Fig. 7-11, transistor Q is represented by its junction diagram while Fig. 7-12 shows the same circuit with the schematic transistor symbol (note the arrowhead on the emitter lead, indicating the direction of current flow). The circuit has a second loop containing the load resistance R_L, and the base-collector junction. To trace

71

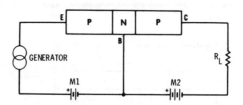

Fig. 7-11. Common-base transistor circuit.

these loops out more definitely we find the input or emitter-base loop beginning at the base B, through battery M1 and generator G, to emitter E and back to base B. From base B to collector C, through load resistance R_L and battery M2, and back to base B is the output or base-collector loop. When the signal voltage is

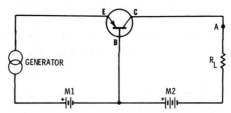

Fig. 7-12. Schematic of common-base transistor circuit.

positive it opposes the forward biasing voltage of battery M1 and reduces the base-emitter current. In turn the collector current is also reduced and the upper end (A) of load resistance R_L becomes more positive. Conversely, as the signal voltage is negative the base current and the collector current increase to make the upper end of load resistance R_L become more negative. Thus, the output voltage developed across R_L is in phase with the signal voltage produced by generator G. However, if the signal voltage has a maximum of 0.01 volt and is impeded by the low resistance—usually about 25 ohms—across the emitter-base junction, the resulting signal current will be 0.4 mA. The alternating component of the collector current may be 25 times the base signal current, or 10 mA. Passing through the load resistance R_L of 100 ohms, this alternating collector current develops an output voltage of 1.0 volt. Therefore, the action of the transistor has in effect increased or amplified the signal voltage from 0.01 to 1.0 volt.

In the common-emitter circuit shown by Fig. 7-13 the signal is introduced into the base-emitter loop. This circuit functions much the same as the common-base circuit but it has greater amplification ability and produces a 180° phase shift between the input voltage and the output voltage. That is, when the signal voltage is negative

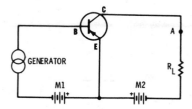

Fig. 7-13. Common-emitter circuit.

the output voltage is positive. Possibly because the common-emitter circuit is more like the triode vacuum tube circuit it is used more often than either of the other two. However, the common-emitter circuit also has the advantage of larger power gains, greater amplifications, and comparatively equal input and output impedances.

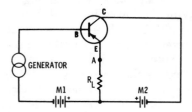

Fig. 7-14. Common-collector circuit.

In the common-collector circuit, the signal is introduced in the base-collector loop, as seen in Fig. 7-14. Although there is no phase shift between the input and the output voltages, the power gain is low and the amplification is less than one for this circuit. However, its input impedance is high and its output impedance is low. Thus, the common-collector circuit is used primarily as an impedance-matching device.

Characteristics

Since the transistor is used most often in the common-emitter circuit the characteristics involving its output element, the collector, are most significant. That is we are primarily interested in the variations of the collector current. Collector current is dependent upon the base current, I_B, as well as that voltage applied between the emitter and the collector. However, the base current in turn is dependent upon the voltage applied across the base-emitter junction.

There are two basic characteristics in which we are interested—*transfer characteristics* and *collector characteristics*. Fig. 7-15 shows the graphic representation of the transfer characteristics of a typical transistor. By holding the collector-to-emitter voltage V_{CE} at a constant value, and varying the base-to-emitter voltage V_{BE} the variation of collector current can be noted. It will be seen that the collector current does not begin until the base-to-emitter voltage reaches a value of 0.6 volt at point (A). This 0.6 volt is the value

73

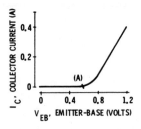

Fig. 7-15. Transfer characteristics of a bipolar transistor.

of the barrier charge normally found across the base-emitter junction and must be overcome before the base or collector currents can start. A similar procedure is followed in obtaining the collector characteristic curves (Fig. 7-16) by holding the base current at a constant value and varying the collector-to-emitter voltage V_{CE}. With the lower levels of base currents the collector current rises rapidly until V_{CE} reaches about 4.0 volts, and then the collector current I_C becomes fairly constant with larger values of V_{CE}.

Quite often in designing circuits that use transistors we are interested in the current gain obtainable between the input and the output. In the common-base circuit, for example, the input current is the emitter current and the collector current is the output current. This current gain is also known as the *forward current-transfer ratio* or *alpha* (α). Current gain for the common-emitter circuit is similarly called the *forward current-transfer ratio* but since it differs from that of the common-base circuit, it is referred to as *beta* (β).

Alpha always has a value of less than one while beta may have a value as great as 100. While beta is a fairly direct indication of the amplification possible with a given transistor, the common-base circuit uses transfer of impedance for its basis of amplification and the alpha factor is not too indicative of the amplification possibility.

That frequency at which the power delivered to the collector load is reduced to half—3 dB down—is referred to as both the *beta cutoff frequency* and the *emitter cutoff frequency*. In general, transistors are not considered usable above this beta cutoff frequency and the various structures described above are used to obtain different beta cutoff frequencies. A transistor having a lower cutoff frequency

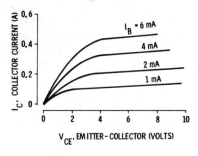

Fig. 7-16. Collector characteristics of a bipolar transistor.

of about 400 kHz is limited primarily to use in audio amplification; one with an upper limit of 2000 MHz permits usage in nearly all phases of electronics.

As with all devices carrying current, heat is developed within a transistor. While the vast majority of transistors are used in low-current circuits (voltage amplification), there are transistors developing as much as 100 watts output and an equal 100 watts of undesired heat. Such a large amount of heat, unless radiated away, will damage the transistor crystal. Several of the transistor structures were specifically designed to function with heat sinks—metal radiators of heat—to dissipate excess heat as quickly as possible. However, each transistor is rated with regard to the amount of power that can be safely handled with or without a heat sink.

UNIPOLAR TRANSISTORS

Considerably newer than the bipolar transistors just discussed, the unipolar, or field-effect, transistor (often abbreviated FET) uses an electrostatic field as a controlling factor. The original FET made use of a pn junction as a *gate* between a *source* and a *drain*. It was much the same as if a garden hose or *source* were placed into one end of a ditch or channel having an outlet or *drain* at the other end. Then if a dam or *gate* is set at some point along that channel the water flow may be controlled. In the FET the gate takes the form of an area depleted of current carriers. Thus, if the channel is made of n-type material and the gate terminal is made negative with respect to the p-type substrate, the channel is depleted of electrons (Fig. 7-17). With the channel depleted of its electron current carriers, the electrons entering the channel at the source cannot travel to the drain.

Since the early type FET used a pn junction as the gate it was essential that it be reverse biased. In this reverse-biased condition the gate presents a very high impedance to any applied voltage, and the power input is almost nothing while the power output is

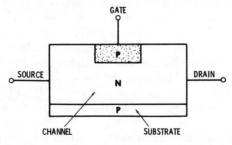

Fig. 7-17. Basic construction of an FET.

comparatively large. However, when the junction gate is forward biased there is considerable input current and the power gain becomes very low. Accordingly the junction-type FET is severely affected by the voltage polarity applied to the gate. To overcome this disadvantage, an insulating layer is now being placed between the metal gate terminal and the channel crystal section, as shown in Fig. 7-18. Since the insulating layer is commonly a metal-oxide,

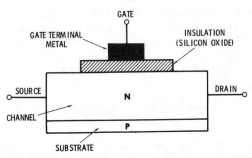

Fig. 7-18. Basic construction of a depletion type MOSFET.

the unit in Fig. 7-18 is referred to as a metal-oxide-semiconductor field-effect transistor, or MOSFET. To further complicate things MOSFETs are also identified as depletion or enhancement types and n-channel or p-channel types. As might be expected the n-channel or p-channel distinction establishes the forward-biased and the reverse-biased conditions. The depletion-type MOSFET conducts current regardless of gate polarity, while the enhancement-type MOSFET conducts channel current only when the gate is forward biased.

Structures

Fig. 7-18 shows the basic construction of a depletion-type MOSFET having an n-channel. Initially, this construction starts with the lightly doped p-type silicon crystal that forms the base or substrate. This substrate is sometimes referred to as the active bulk and is often marked with a B in the schematic symbols (see Fig. 7-19), but is not considered to be an active element of the transistor. Steps of masking, acid etching, and diffusion, etc., build the channel, the metal contacts for the source, the drain and the gate, as well as the silicon-dioxide insulation between the gate terminal and the channel. If this were to be a p-channel MOSFET the crystal types would be reversed with the substrate being n-type crystal material.

Structure of the enhancement-type MOSFET, as seen in Fig. 7-20, differs from the depletion type by having two channel sec-

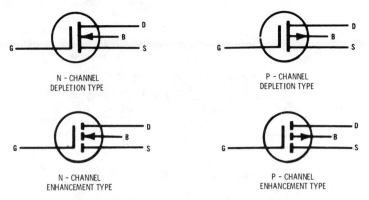

N - CHANNEL
DEPLETION TYPE

P - CHANNEL
DEPLETION TYPE

N - CHANNEL
ENHANCEMENT TYPE

P - CHANNEL
ENHANCEMENT TYPE

Fig. 7-19. Schematic symbols for MOSFET.

tions divided by a portion of the substrate. The gate region covers portions of both the source and the drain as well as that portion of the substrate between the two channel sections. When enough positive voltage is applied to the gate of the n-channel MOSFET that portion of substrate between the two channel sections changes from p-type to n-type. The channel is then "open" to provide a path for the conduction of current. Actual construction of the enhancement-type MOSFET follows the same masking, acid-etching and diffusion procedure as that for the depletion-type MOSFET.

Fig. 7-21 shows the construction of a dual-gate MOSFET having two channel sections covered by two metal gate terminals. Between the two channel sections is a block of channel material that is described as the drain of the first channel and the source of the second. However, this middle block (indicated by X in Fig. 7-21) has no external connection and little effect. Basically, the voltage applied to each gate acts upon the current flowing from the source to the drain. Usable as rf amplifiers, gain-controlled amplifiers, mixers, modulators, demodulators, etc., the dual-gate MOSFET is very versatile. The gain-controlled amplifier, having an rf signal applied to the first gate and a dc voltage on the second gate, is

Fig. 7-20. Structure of the enhancement-type MOSFET.

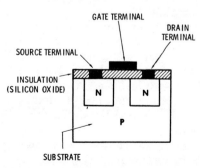

77

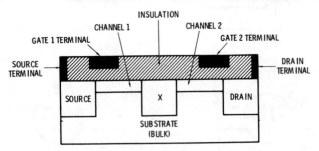

Fig. 7-21. Construction of dual-gate MOSFET.

similar to the reactance transistor portion of the voltage controlled oscillator (vco) discussed in a later chapter.

Basic Circuits

Much like the bipolar transistor, the FET has three basic circuit forms—the common-source circuit, the common-gate circuit, and the common-drain circuit. Most frequently used of these basic circuits is the common-source arrangement; it has a high input impedance, medium to high output impedance, and greater than unity voltage gain. Its input signal is applied between the gate and the source, while the output signal is developed across a load resistor, R_L, between the drain and the source, as indicated in Fig. 7-22. There is a 180° phase shift between the input and the output of this circuit.

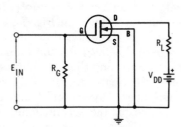

Fig. 7-22. Common-source FET circuit.

Very high input impedance, low output impedance, a voltage gain of less than unity, and no phase shift between the input and the output are among the characteristics of the common-drain arrangement shown in Fig. 7-23. This common-drain circuit is also referred to as a *source-follower*. With the input signal injected between the gate and the drain and the output coming from between the source and the drain, the circuit has a very low input capacitance.

Fig. 7-24 shows the common-gate circuit used to transform a low input impedance to a high output impedance. The output

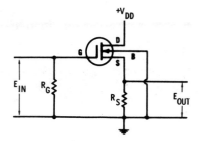

Fig. 7-23. Common-drain FET circuit.

voltage developed between the drain and the gate is 180° out-of-phase with the input voltage applied between the source and the gate. Voltage gain for this circuit is greater than unity but not greater than 2 in most instances.

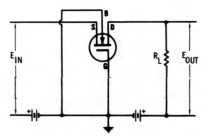

Fig. 7-24. Common-gate FET circuit.

Characteristics

To learn the characteristics of a MOSFET a common-source circuit like that seen in Fig. 7-25 is necessary. With a fixed voltage between the source and the gate, various voltages are applied between the drain and the source, and a notation made of the drain current. Typically, with −1.0 volt between the source and the gate and +1.0 volt between the drain and the source the drain current might be 2 mA, as shown at point A of Fig. 7-26. At this low level of drain current the channel is unrestricted except for its ohmic resistance and if the drain-to-source voltage (meter M2) is doubled

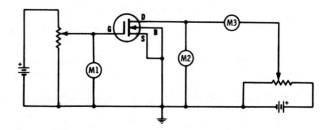

Fig. 7-25. Circuit for determining characteristics of a MOSFET.

to +2.0 volts, the drain current is also doubled to 4 mA. However, further increase of the drain-to-source voltage creates quite a voltage differential between the gate and the drain to cause a partial depletion of current carriers within the channel. Thus, the drain current increases at a much slower rate beyond point B (Fig. 7-26). And, beyond point C, there is little or no change in drain current with increases in drain-to-source voltage.

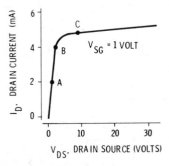

Fig. 7-26. Experimental drain characteristics for a MOSFET.

With other values of source-to-gate voltage V_G, the curves of Fig. 7-27 are the output characteristics of the MOSFET. Figure 7-27A shows the characteristics of a depletion-type MOSFET while the enhancement-type MOSFET characteristics are shown by Fig. 7-27B. It should be noted that the depletion type has gate voltage values of both positive and negative polarities, while the enhancement type has gate voltages with only positive polarities. The characteristics shown are for an n-channel MOSFET; the gate voltages would be of opposite polarities for p-channel transistors. Fig. 7-28 shows the transfer characteristics for a MOSFET.

In designing circuits using MOSFETs an important factor is the forward transconductance Y_{FS}. This corresponds to the mutual or transconductance of the electron tube, and is a measure of effect of the gate voltage, V_G, upon the drain current, I_D.

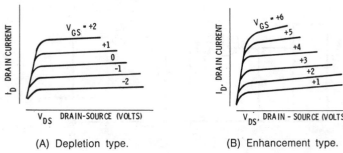

(A) Depletion type.   (B) Enhancement type.

Fig. 7-27. Typical output characteristics for a MOSFET.

$$Y_{FS} = I_D/V_G \text{ (for direct current)}$$

$$Y_{FS} = \frac{I_{D2} - I_{D1}}{V_{G2} - V_{G1}} \text{(for alternating current)}$$

where,
I_{D2} is the drain current when the gate voltage is V_{G2},
I_{D1} is the drain current when the gate voltage is V_{G1}.

Thus, the MOSFET can be considered to be a constant current device having an output equaling the gate voltage V_G multiplied by the forward transconductance Y_{FS}.

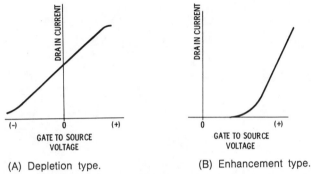

(A) Depletion type.   (B) Enhancement type.

Fig. 7-28. Transfer characteristics for a MOSFET.

REVIEW QUESTIONS

1. What is the difference between the bipolar and the unipolar transistor? What are the three crystal sections of a bipolar transistor?
2. Name the three currents present within a working bipolar transistor. If the collector current equals 1.0 mA and the base current is 0.02 mA, what is the emitter current?
3. Describe the process of forming an alloy-junction–type transistor. What is the primary advantage of the drift field transistor?
4. What is the primary difference between the homotaxial transistor and the epitaxial transistor?
5. Name the three basic transistor circuit forms. Which of these common circuit forms is most commonly used? Is there any phase shift between the input and the output of the common-emitter circuit?
6. Name the two basic characteristics for the bipolar transistor. What is the forward current-transfer ratio for the common-base circuit? For the common-emitter circuit?

7. Name the three parts of an FET.
8. Name one disadvantage of the junction-type FET. How does the MOSFET differ from the junction-type FET? How does the depletion-type MOSFET differ from the enhancement type?
9. Is the substrate an active part of the MOSFET? Does the enhancement-type transistor have one or two channel sections?
10. Name the three basic circuit forms for the FET.
11. What is the forward transconductance of an FET? If the signal voltage is 1.0 volt, the Y_{FS} is 10,000 μmhos, and the load resistance R_L is 5,000 ohms, what is the output voltage?

CHAPTER 8

Integrated Circuits

Serving as resistances and capacitances as well as transistors and diodes, solid-state devices provide a means of combining many electronic components into a single unit. Separately the individual transistors, diodes, resistors, inductors, capacitors, etc., are called discrete components. Combined into a single unit—usually on a chip of silicon—the many components become an *integrated circuit*. One of the simpler integrated circuits contains two zener diodes, five diodes, eleven transistors and five resistors on a silicon chip measuring 0.05 inch (about half the distance between two periods .. on a standard typewriter) on each side. Fig. 8-1 gives an enlarged view of an IC chip for a voltage regulator, and Fig. 8-2 compares the size of several such chips with the common quarter (25¢ coin). Integrated circuits have given the electronics field many new dimensions—not only in the sense of actual space but in areas previously considered entirely foreign to electronics.

BASIC PRINCIPLES

In electronics there are a number of circuits that are used repeatedly. One prime example of such a circuit is the resistance-capacitance coupled audio voltage amplifier shown in Fig. 8-3. In the final analysis this circuit consists of a transistor, 3 resistors, and one capacitor. As the heart of the circuit, the transistor has a cross-section as shown by Fig. 8-4A. Note that this transistor is constructed with its three contacts on top. The top of that transistor can be constructed so as to have an appearance as that shown by Fig. 8-4B.

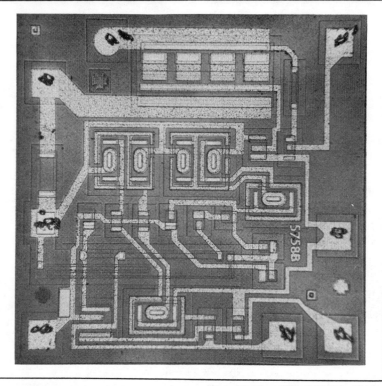

Courtesy RCA, Solid-State Division

Fig. 8-1. Enlarged view of an IC chip for a voltage regulator.

Crystal silicon is a resistive material and its resistivity can be altered with differing concentrations of impurities. That is, pure silicon is highly resistive, while the addition of impurities and the resulting free electrons or holes decrease its resistivity. If a strip of p-type silicon is arranged between two metal contacts as shown in Fig. 8-5A, it will serve as a resistor. The p-type substrate and the n-type sections serve primarily for physical reasons. Values of resistors formed in this manner range from about 1 to 20,000 ohms, with tolerances of better than 10%. Fig. 8-5B shows the top of this resistor.

Capacitors can be formed in two different ways. Fig. 8-6A shows a capacitor using the capacitance characteristic of a reverse biased pn junction. Naturally such a capacitor will not function properly if the voltage applied is of the wrong polarity or alternating. Somewhat more like a conventional capacitor, that shown in Fig. 8-6B uses an n-type section for one plate and a layer of silicon oxide for its dielectric. The other plate is a layer of metal deposited at

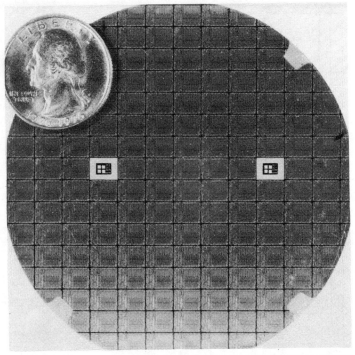

Courtesy RCA, Solid-State Division

Fig. 8-2. Silicon wafer containing several IC chips compared with the size of a quarter.

Fig. 8-3. Transistor RC coupled audio amplifier.

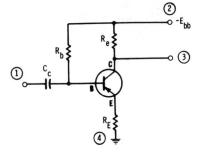

the same time as the interconnecting pattern. Neither of these capacitor forms can be constructed with values above a few hundred picofarads. Fig. 8-6C shows what might be the top of either of these capacitor types.

Depths of the transistor in Fig. 8-4A, of the resistor in Fig. 8-5A, and of the capacitor in Fig. 8-6A or 8-6B can all be made equal.

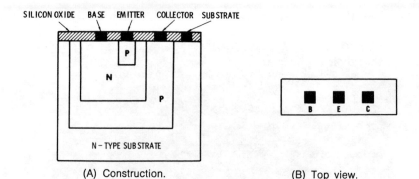

(A) Construction.

(B) Top view.

Fig. 8-4. Integrated circuit transistor.

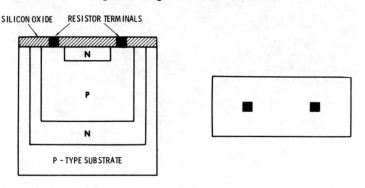

(A) Construction.

(B) Top view.

Fig. 8-5. Integrated circuit resistor.

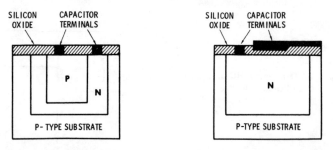

(A) Junction capacitor construction.

(B) Metal-oxide capacitor construction.

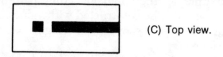

(C) Top view.

Fig. 8-6. Integrated circuit capacitor.

Therefore, the resistor unit, R_E can be placed so that one of its contacts is adjacent to the emitter contact of the transistor Q unit (Fig. 8-7). In Fig. 8-8, capacitor C_C and resistor R_B have been set so that their contacts are near the base contact of transistor Q.

Fig. 8-7. Positions of resistor R_E and transistor Q.

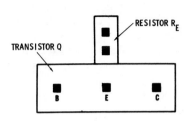

Fig. 8-8. Addition of C_C and R_B units.

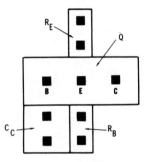

Fig. 8-9. Addition of resistor R_C.

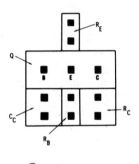

Fig. 8-10. Completed integrated circuit with metal connections.

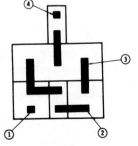

Close to the collector contact, the one contact of resistor unit R_C has its other contact near the second contact of the R_B unit (Fig. 8-9). With metal connections between these different units (Fig. 8-10), the circuit of Fig. 8-3 has been completed. But each of the units—transistor Q, resistors R_E, R_B, and R_C, and capacitor C_C—are "encased" within a p-type substrate and can be constructed upon a single slice of p-type silicon. That is, by repeating a process of oxidizing, masking, and diffusing or growing crystal sections the various transistors, diodes, resistors, and capacitors can be formed into a single unit.

While these above paragraphs and illustrations are fairly accurate on the basic principles of integrated circuits, the actual integrated circuit, IC, is very different. In practical ICs it is impossible to build many of the resistors or capacitors needed in circuits like the RC coupled audio amplifier—the chip space required for such resistors or capacitors is often prohibitive.

COMPONENT ALTERATIONS

Although very few of us will ever design an IC, the rules followed by designers assist greatly in understanding their functioning. These design rules are:

1. Reduce the number of leads coming from an IC. A major cost of producing an IC is the attaching of leads. Circuit performance is also often improved by reducing the number of leads and the amount of external circuitry.
2. No inductors. Inductors, as we know them, have length, width and depth while the IC chip is basically a single plane structure with little or no depth. And because of size even external inductors are used infrequently.
3. Reduce total capacitance. Capacitance is directly related to a capacitor's plate area, which means the IC chip area. And any reduction in chip area also reduces the cost.
4. Reduce total resistance. Resistance is directly related to length of current path which, in turn, means chip area. Therefore, resistance, especially of large resistors—100,000 ohms or greater—is a costly factor.
5. Use as many matching components as possible. Tolerances on IC components are commonly very poor. For example, an IC resistor may vary as much as 20%—a 1000-ohm resistor might have actual values ranging from 800 to 1200 ohms. However, two resistors fabricated on the same chip will likely have matching values. Although the 1000-ohm resistor may have an actual value of 920 ohms, all similarly designed resistors

on that chip will likely have actual values of 920 ohms. Transistors, diodes, etc., of similar design on the same chip will also be apt to match. In other situations, resistors—or other components—having equal factors of error are advantageous. Resistors designed to have values of 500 and 1000 ohms and having actual values of 600 and 1200 ohms will function better than separate resistors of 600 and 800 ohms. That is, the ratios of 500/1000 and 600/1200 are equal but 500/1000 and 600/800 are not equal.

6. Substitute transistors and diodes for larger area components. Fig. 8-11 shows a combination of two transistors and two small resistors which uses only about $\frac{1}{10}$ the chip area re-

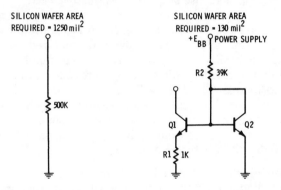

Fig. 8-11. Area comparison for circuit alternatives.

quired by a single 500,000-ohm resistor. Transistor Q1 serves as a diode biasing transistor Q2 to permit a given collector current. Typically the 0.7-volt bias developed across Q1 establishes a 0.7 mA through the base-emitter of Q2 and 20 mA through its emitter-collector circuit. The combination of Q1, Q2, R1 and R2 passes a constant current much the same as the 500,000-ohm resistor of Fig. 8-11.

7. Use built-in voltage references. A voltage reference provides a constant voltage independent of the supply voltage—the 117-volt ac from the electric company can vary as much as 10% and batteries can vary even more. A voltage reference overcomes these supply voltage variations by either passing a greater amount of current until a series resistor drops the voltage to a known value or by having a constant voltage drop regardless of the current passed. Fig. 8-12A shows a reverse-biased diode acting as a voltage reference across a load resistor R_L. Operating at or near its reverse-breakdown voltage, diode D1 conducts current when the load voltage exceeds

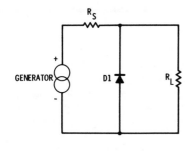
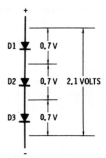

(A) Reverse-biased diode used as voltage reference.

(B) Series-connected diode used as voltage reference.

Fig. 8-12. Use of diode as voltage reference.

that reverse breakdown voltage. In turn, this produces a larger voltage drop across series resistor R_S and reduces the load voltage E_L. Similarly, a decrease in the supply voltage E_o produces a smaller voltage drop across R_S while diode D1 acts to increase the load voltage E_L.

Forward-biased silicon diodes, passing current, have a constant voltage drop between terminals of about 0.7 volt. Therefore, the three silicon diodes in Fig. 8-12B have fixed voltage of 2.1 (3×0.7) between A and B.

Summarizing these rules tells us that the IC uses several transistors and diodes that are actually auxiliary to the primary circuit. In one typical IC amplifier less than half of the 20 transistors are used for amplification—the other transistors serve as constant current sources, voltage references (base-emitter junctions serving as silicon diodes) and level shifters.

CONSTANT-CURRENT SOURCES

A *constant-current source* is any generator or other device providing a fixed level of current flow regardless of its load. Fig. 8-13 shows a generator G_{CC} producing a constant current of 10 mA to a load resistor R_L. Theoretically, R_L can be of any value while the current will always be 10 mA. In ICs constant-current sources are

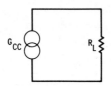

Fig. 8-13. Constant-current generator and resistance load.

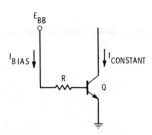

Fig. 8-14. Constant-current source using a transistor and resistor.

not generators but instead are *current sinks* passing a given rate of current. To some extent, any resistor is a current sink. However, since it is not practical to use large resistors in ICs, transistors are used very effectively as substitutes for resistors and current sinks. A very simple constant-current source is shown in Fig. 8-14 using a transistor and a resistor as a current sink, based upon the assumption that the transistor collector-emitter current I_c equals the base-emitter current I_b times the current gain β. Assuming that the supply voltage E_{BB} is constant, I_b will be constant and β is constant so that I_c will also be constant.

Actually the resistor used in the current sink of Fig. 8-14 is too large and the current gain is seldom as constant as desired. To correct these disadvantages, the circuit of Fig. 8-15 uses two well-matched transistors. Q2, as a forward-biased diode, acts as a voltage reference. And although the collector current of Q2 has little effect upon the circuit, it does increase the voltage drop across resistor R and permits R to be of a smaller value. Q1 is thus biased

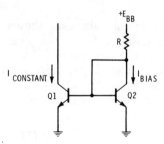

Fig. 8-15. Improved constant-current source.

by the voltage across the base-emitter of Q2 and permits a constant Q1 collector-emitter current. For the sake of simplicity, many of the constant current sources in complex ICs will be shown as a generator G_{cc}.

VOLTAGE REFERENCES

As discussed previously, voltage references are of two basic types —a forward-biased diode or a reverse-biased diode. Because of

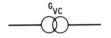

Fig. 8-16. Generator symbol to be used in place of a voltage reference.

the reduction in the error factor, well-matched transistors—their base-emitter junctions—are often used in place of forward-biased diodes. Whenever possible a voltage reference will be replaced by a generator symbol identified as G_{vc} (Fig. 8-16).

LEVEL-SHIFTER CIRCUITS

Since most solid-state circuits, including those in ICs, are direct coupled, the level of direct current and dc voltage becomes excessive. Correction of this excessive dc level is accomplished by a level-

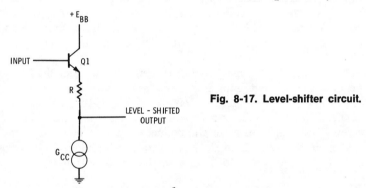

Fig. 8-17. Level-shifter circuit.

shifter circuit like that shown in Fig. 8-17. With the constant-current source G_{cc} developing constant voltage drops across the base-emitter junction of Q1 and R, the output is at a lower dc level. Such a lower dc level substantially reduces the power consumed by the output amplifier—often zero when there is no input signal from previous stages of the IC.

OUTPUT STAGES

Output stages of ICs deliver differing amounts of power—audio amplifier ICs for smaller radios and television receivers drive the speaker directly while larger audio equipment requires another power stage. This is also true for other frequency ranges but the audio range is more common to the use of ICs. In either case, most linear ICs include a power amplifier as the output stage. Such power amplifier stages are of two basic types—single-ended or npn/pnp complementary as shown by Fig. 8-18. The single-ended power amplifier (Fig. 8-18A) has the disadvantage of requiring a biasing

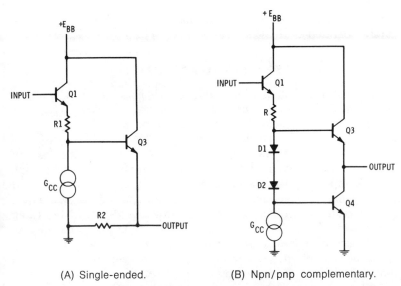

(A) Single-ended.   (B) Npn/pnp complementary.

Fig. 8-18. Power output amplifiers.

current, while the npn/pnp complementary draws no current (power) when no signal is applied. In the schematic drawing of Fig. 8-15B, Q3 and Q4 are power amplifier transistors while Q1, G_{cc} (current source), R, D1 and D2 form a level shifter. Deriving its input across diodes D1 and D2, power is developed as Q3 and Q4 conduct large currents. Because of the complementary characteristics of the npn transistor (Q3) and the pnp transistor (Q4), the circuit produces the results of a push-pull amplifier without an inverting circuit. High-current pnp-type transistors are often rather costly and a smaller low-current pnp-type is used with a high-current npn-type as a means of effecting a high-current pnp-type—a quasi-complementary output amplifier.

There are, as we will see in following chapters, a great many other circuits common to ICs. However, many of those circuits are related to either linear ICs or digital ICs.

ELECTRICAL CHARACTERISTICS

The electrical characteristics of an IC are much the same as those of any electrical device. Quite often these characteristics are of importance only when the IC is used in a specified manner. While linear ICs are rated by amplification and frequency range (bandwidth), digital ICs have little relation to these factors. Among those factors related to all ICs are input level, input impedance, output level, output impedance, power dissipation, supply current, supply voltage, and temperature (maximum at which IC may be safely operated).

BLACK-BOX APPROACH

To a great extent, the integrated circuit is a sealed "black box" with only the manufacturer's specifications to follow. Thus, we are basically interested in knowing the resulting effect of putting a signal into an IC—whether the signal will be amplified, distorted or otherwise altered in shape, etc. The IC input is the load of the generator developing the signal—a microphone, a phonograph pickup, tape playback head or another IC stage. The output of an IC feeds into another IC, a loudspeaker, a digital readout, a printer readout or other device converting electrical variations into forms which are visible or audible. The important factors are to match the IC characteristics with the input and output devices. Actually the internal circuits of the IC are important only as a matter of better understanding the final result and make outboard circuitry (those connections and components outside the IC) serve the desired purpose. For example, the operational-amplifier IC com-

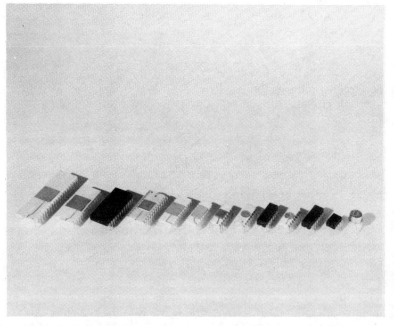

Fig. 8-19. IC package styles.

monly has a noninverting input and an inverting input with differentiating amplifiers to provide a multitude of possibilities. As will be shown in a later chapter, such an operational-amplifier (op-amp) IC provides for distortion-free high-gain audio amplification among other things. A number of IC package styles are shown in Fig. 8-19.

REVIEW QUESTIONS

1. What are discrete components? What is an integrated circuit?
2. Describe in simple terms, a solid-state resistor. A solid-state capacitor.
3. Name seven rules of design followed in the production of ICs.
4. What is a constant-current source? What is a current sink?
5. What is a voltage reference?
6. Why does the dc level of solid-state circuits become excessive? What type circuit is used to correct this situation?
7. Name one advantage of the npn/pnp complementary power amplifier.
8. List eight factors or characteristics related to all ICs.
9. Is it absolutely essential to understand the internal circuits of an IC?

CHAPTER 9

Basic Amplifier Circuits

The purpose of an amplifier is to increase the amplitude of a voltage or a current. An amplifier is a transistor or integrated circuit creating a voltage, current or power having an amplitude larger than the controlling factor. The controlling factor—usually a small voltage or current—may have a number of different forms or frequencies. In electron tube amplifiers there was considerable concern with the frequency range of an amplifier as capacitance and inductance changed their reactance with frequency. Amplifiers using transistors or integrated circuits use few, if any, capacitors and inductors and are, therefore, less affected by frequency.

TRANSISTOR CIRCUIT

When a bipolar transistor is the primary component of an amplifier circuit, the base-emitter current is the controlling factor. This base-emitter current reduces the normally high resistance of the emitter-to-collector path. A base-emitter current decrease produces a large decrease in the emitter-collector current. Conversely, an increase in the emitter-collector current results from an increase in the base-emitter current.

As shown in Chapter 7, the bipolar or junction transistor functions as a junction, with the emitter current equaling the total of the base and collector currents. The collector current I_c is equal to the base current I_b times the forward current transfer ratio β. If the applied or signal current $I_b{}'$ flows through the emitter-base junction, the emitter-collector current $I_c{}'$ equals $\beta I_b{}'$.

$$I_c{}' = \beta I_b{}'$$

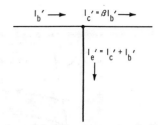

Fig. 9-1. Junction equivalent of bipolar transistor fundamentals.

Fig. 9-1 shows the junction equivalent of these transistor fundamentals.

Fig. 9-2A shows a basic transistor amplifier circuit while the base-emitter equivalent is shown by Fig. 9-2B. This base-emitter equivalent is a series circuit of biasing battery M1, signal source G, R_{eb} between the emitter-base junction, and emitter resistor R_E. Battery M1 produces biasing current I_b. Signal source G increases

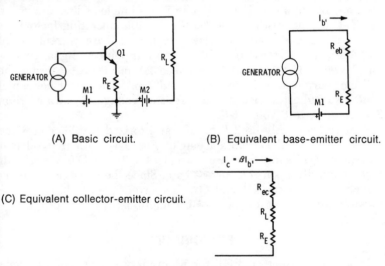

(A) Basic circuit.   (B) Equivalent base-emitter circuit.

(C) Equivalent collector-emitter circuit.

Fig. 9-2. Transistor amplifier.

and decreases the base current by an amount equaling I_b'. In turn, the collector current I_c' varies by an amount equaling $\beta I_b'$. This collector current is shown acting upon the equivalent series circuit in Fig. 9-2C of emitter-collector resistance R_{ec}, load resistor R_L, and emitter resistor R_E. Output voltage E_o is thus developed across R_L equaling $\beta I_b' R_L$. Recognizing that both the base current I_b' and the collector current I_c' (or $\beta I_b'$) passes through emitter resistor R_E, the input voltage E_i and the voltage amplification can be determined.

Voltage amplification, $E_o/E_i = R_L/R_E$

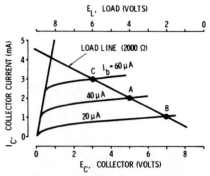

Fig. 9-3. Graphic representation of transistor amplifier action.

A graphic solution or representation of this amplification can be seen on the graph of Fig. 9-3. The 2000-ohm load line drawn on the E_c-I_c characteristics indicates the possible operating factors. As shown the circuit could be biased to have a base current, I_b, of 40 μA and a collector current, I_c, of 2 mA at point A. Signal generator G swings base current I_b' between 20 and 60 μA causing the collector current to vary between 1 and 3 mA. The upper horizontal scale also indicates that the load-resistor voltage E_L swings from 2 volts, at B, to 6 volts at C.

A more commonly used form of transistorized amplifier uses the directly connected Darlington pair of Fig. 9-4. With the collector current of Q1 being the base current of Q2, the Darlington pair relies primarily upon the current gain. Since there are no coupling components between Q1 and Q2, the current gain is basically the forward current transfer ratio of Q1 multiplied by that of Q2.

FET CIRCUIT

In a circuit using an FET for its primary component, the gate voltage is the amplifier controlling factor. In effect, the gate voltage regulates the current flow between the source and the drain. Fig.

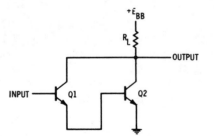

Fig. 9-4. Darlington pair circuit.

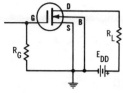

| (A) Basic circuit. | (B) Equivalent drain circuit. |

Fig. 9-5. FET amplifier.

9-5A shows the actual FET amplifier circuit while the equivalent source-drain circuit is indicated by Fig. 9-5B. The drain current equals the gate voltage V_{sg} times the forward transconductance Y_{FS}.

$$\text{Drain current, } I_D = Y_{FS}V_{sg}$$

The output voltage E_o is then the product of I_D times the load resistance R_L.

$$\text{Output voltage, } E_o = I_D R_L$$
$$= Y_{FS}V_{sg}R_L$$

Of course, the gate voltage V_{sg} is also the input voltage E_i and the voltage amplification is $Y_{SF}R_L$.

$$\text{Voltage amplification, } E_o/E_i = Y_{FS}R_L$$

Typically, the forward transconductance Y_{FS} is 10,000 μmho and with a load resistor R_L of 5000 ohms, the voltage amplification is 50.

REVIEW QUESTIONS

1. What is the purpose of an amplifier?
2. What is the controlling factor of a bipolar transistor amplifier?
3. Describe briefly the amplification process of the transistor amplifier. If a transistor amplifier has a β of 50 and an R_L of 5000 ohms and an R_E of 250 ohms, what is the voltage amplification?
4. Describe the Darlington pair. What is the current gain of a Darlington pair, if each transistor has a β of 50?
5. What is the voltage amplification of an FET amplifier having a forward transconductance of 15,000 μmho and a load resistor of 2000 ohms?

CHAPTER 10

Operational Amplifiers

An operational amplifier (usually abbreviated as op amp) is very basically defined as a very-high gain dc amplifier. However, any circuit that will successfully amplify dc will also amplify ac to a rather high frequency.

OP-AMP BASICS

Initially op amps were heavy, costly and bulky electron tube amplifiers and while transistors and other discrete solid-state components made op amps a bit more practical, the IC op amp is the biggest advance. Actually it is now difficult to imagine the complex construction of an op amp with discrete components. The IC provides versatility and reliability that could never be attained with discrete components.

Commonly the op amp has dual channels of amplification—one channel, with an input marked by a +, is noninverting and the other is inverting (−). That is, one channel changes (inverts) a positive signal to a negative signal. Eventually, the two channels feed into one output. So if the two inputs are acted upon by the same signal, see Fig. 10-1, the resultant output will be zero—the two amplified signals being 180° out of phase. When one input signal is slightly larger than the other, the output equals the amplified difference. That is, if the signal applied to the +input is 5 volts and that to the −input is 4.5 volts, the output will be 0.5 (5 − 4.5) volts times the amplification factor. The portion of the two signals not amplified—4.5 volts in this example—is referred to as the *common-mode signal*. And the op amp is said to be a differ-

Fig. 10-1. Symbolic representation of IC op amp.

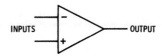

ential amplifier having *common-mode rejection*. In our example, the difference between the two signals—0.5 volt—is referred to as the *differential input voltage*. It may be wondered why common-mode rejection would be desired. One very basic need is the rejection of 60-Hz noise which becomes an annoyance in many electronic applications. In audio equipment, the 60-Hz noise is readily recognized as a hum. The very small signals encountered in medical usage of electronic equipment (electrocardioscopes, etc.) can be lost amid 60-Hz noise unless common-mode rejection is applied.

As explained in Chapter 8, it is somewhat useless to describe the internal circuits of an IC since they are fixed and very complex. However, there is also considerable flexibility by way of secondary input and output terminals on the typical op amp. For our studies we shall discuss only circuits peculiar to op amps and their segments.

Differential Amplifiers

Fig. 10-2 shows a block diagram of a typical op amp. The first stage of nearly all op amps is a *differential amplifier*. Differential amplifiers amplify the difference between two signals. Fig. 10-3

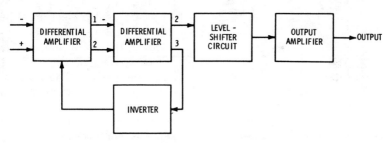

Fig. 10-2. Block diagram of a typical op amp.

shows the circuit of a differential amplifier having the emitters of Q1 and Q2 fed by a constant-current source G_{cc}. In this manner the total of Q1 and Q2 base and collector currents is also comparatively constant. An increase in the base current of Q1 produces an increase in Q1 collector current. But an increase in Q1 collector current reduces the current going to the emitter of Q2. In turn, the collector current of Q2 is decreased. That is, if G_{cc} develops 10 mA, it normally divides equally with 5 mA going to each emitter. Increasing the collector current of Q1 to 6 mA "robs" 1.0 mA from

Table 10-1. Conditions of Operations for Differential Amplifier

+ INPUT	− INPUT	OUTPUT 1	OUTPUT 2	OUTPUT 3
1. Grounded	Grounded	Constant dc, no output signal	Constant dc, no output signal	Constant dc voltage, no output signal
2. Small ac signal	Small ac signal	Constant dc, no output signal	Constant dc, no output signal	Ac voltage output signal
3. Distorted ac signal	Distorted ac signal	Constant dc, no output signal	Constant dc, no output signal	Ac voltage output signal, distorted
4. Square-wave signal	Square-wave signal	Constant dc, no output signal	Constant dc, no output signal	Very-short duration pulses, or blips, occurring as signal changes polarity
5. Saw-tooth signal	Sawtooth signal	Constant dc, no output signal	Constant dc, no output signal	Sawtooth signal
6. Small ac signal	Grounded	Amplified ac, inverted	Amplified ac	Constant dc voltage, no output signal
7. Grounded	Small ac signal	Amplified ac	Amplified ac, inverted	Constant dc voltage, no signal output
8. Distorted ac	Grounded	Amplified distorted inverted ac	Amplified distorted ac	Constant dc voltage, no signal output
9. Distorted ac	Small ac signal	Amplified inverted difference	Amplified difference	Voltage equivalent of signal difference
10. Ac signal	Inverted ac signal	Amplified ac, inverted	Amplified ac	Constant dc voltage no signal output
11. Distorted ac signal	Inverted distorted ac signal	Inverted distorted ac	Distorted ac	Voltage variation at distortion

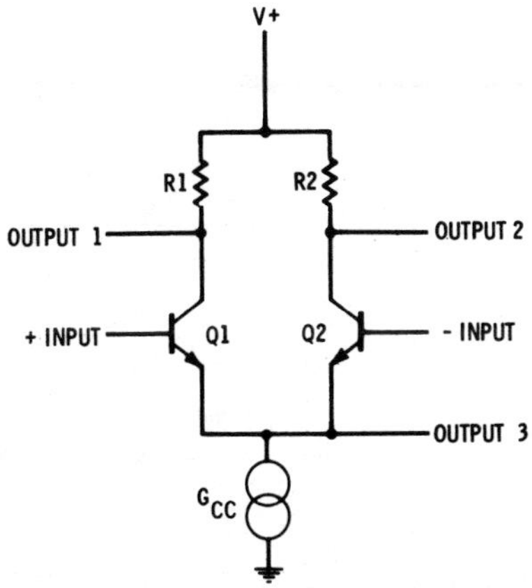

Fig. 10-3. Differential amplifier circuit.

Q2, making Q2 collector current 4 mA (assuming the base currents to be relatively small). The two collector currents, I_{c1} and I_{c2}, total to equal the current of G_{cc}, I_{c3}.

$$I_{c3} = I_{c1} + I_{c2}$$

Understanding the differential amplifier requires studying different input signals applied to the circuit.

To simplify the study of the differential amplifier, Table 10-1 lists a number of input combinations and the resultant outputs. Situations 1, 2, 3, 4 and 5 in Table 10-1 are common-mode signal situations with no signals being developed at either output 1 or output 2. Some of these common-mode situations do develop a signal at the common emitter connection—output 3—that is often used for feedback purposes. By taking an undesirable portion of a wave—its distortion—from output 3, that undesirable portion can become the common-mode signal of another differential amplifier. Then, with the common-mode signal being the distortion, the second differential amplifier eliminates—or reduces—the distortion.

Fig. 10-4 shows the input section of one op amp used basically for audio amplification. An audio signal—small ac signal—is applied to the + input of the first differential amplifier while the − input is grounded. This is situation 6 in Table 10-1 giving an amplified signal on output 1 and an inverted amplified signal on output 2. Fed directly, these push-pull (180° out of phase) signals are doubly amplified by the second differential amplifier—situation 10 in Table 10-1. Output 2 of this second differential amplifier feeds the following op amp stage, while output 3 detects any distortion and feeds back to a common point of supply voltage as a matter of compensation. Such compensation is especially desirable in audio amplification and other situations requiring linear amplification—

amplification of a waveform without changing or distorting its original shape.

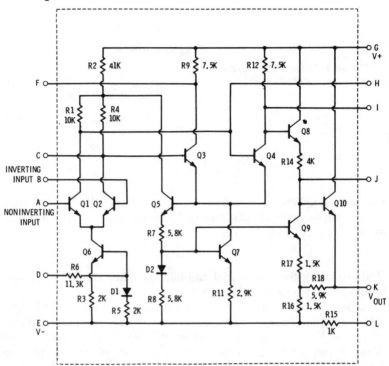

Fig. 10-4. General-purpose op amp such as used for audio amplification.

Inverters

Within the input section of the op amp shown in Fig. 10-4, transistor Q5 serves as an inverter to give the signal from output 3 a 180° phase shift. A positive voltage pulse from output 3 becomes a negative voltage pulse at the collector terminal of Q5. The inverter does not amplify a signal.

Darlington Pair Amplifier

Adding to the current gain, the differential amplifier sections are commonly followed by a Darlington pair (refer to Fig. 9-4). The general-purpose op amp of Fig. 10-4 does not show a Darlington pair—probably because the dual differential amplifier stages are giving sufficient amplification. In effect, Q4 and Q8 in Fig. 10-4 form a Darlington pair even though Q4 is a part of the differential amplifier and Q8 also serves as a portion of the level shifter.

Level-Shifter Circuit

With three stages amplifying the signal by a multiple of at least 50,000, the initial biasing dc of 5.0 μA would be an excessive 250 mA. To shift this excessive current, and related voltage, the combination of Q8, Q9, R14, and R17 serves as a level-shifter circuit.

Output Stages

As shown in Fig. 10-4, the output stage transistor Q10 is of the single-ended type having rather low-output power and draws considerable current when no signal is being applied. Taking the output power from the emitter terminal of Q10 makes this an *emitter-follower circuit* (aside from giving the circuit a type name, this means simply that the same power is available at the emitter as at the collector). If the op amp had been intended to develop larger amounts of power, the npn/pnp complementary circuit of Fig. 8-15 would likely have been used.

OP-AMP USES

An op amp provides very-high gain over a wide range of frequencies, making it useful in many applications. Initially op amps were intended to perform mathematical operations—integration, differentiation, summation and subtraction. Today, op amps are used in nearly every type of electronic device.

Audio Amplifiers

Functioning with little variation from 0 Hz (dc) to above 1.0 MHz, the op amp gives voltage gains of from 1000 to 1,000,000 over the entire audio frequency range (30 to 20,000 Hz). Fig. 10-5

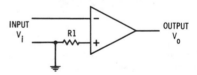

Fig. 10-5. Simple audio op amp.

shows a very simple "straight forward" amplifier using only the −input (the +input floats). The particular op amp used in Fig. 10-5 has gain of about 1000—a commonly desired amount of gain. However, it is often desirable to isolate the input of an audio amplifier from its output. The overall gain—the *open-loop gain*—of an op amp, as used in Fig. 10-5, is the ratio of output voltage V_o to input voltage V_i.

$$\text{Open-loop gain, } G_o = \frac{V_o}{V_i}$$

With feedback, as seen in Fig. 10-6, output voltage V_o will cause a current I_f to flow through the feedback impedance Z_f. I_f also flows through the op amp's input impedance Z_{io} in completing its circuit, or loop. Then by Kirchhoff's law:

$$V_o - I_f Z_f - I_f Z_{io} = 0$$

In the same manner, a signal current I_s results from an application of signal voltage V_s and flows through input component Z_{in} and input impedance Z_{io}.

$$V_s - I_s Z_{in} - I_s Z_{io} = 0$$

Currents I_s, I_f, and I_i flow into the junction at the −input terminal and add to a resulting zero.

$$I_s + I_f + I_i = 0$$
$$\text{or } I_s + I_f = -I_i$$

By comparison, the input voltage V_i($I_f Z_{io}$ or $I_s Z_{io}$) is very small, and I_i is equally insignificant. Then:

$$I_s = -I_f$$
$$\text{Closed-loop gain, } G_c - \frac{V_o}{V_s} = \frac{Z_f}{Z_{in}}$$

Since the signal voltage V_s, in Fig. 10-6, is applied to the −input (inverting), it follows that the feedback current I_f will be negative when the signal current I_s is positive. Thus, the closed-loop gain

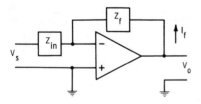

Fig. 10-6. Audio op amp using negative feedback.

is not affected by the open-loop gain, or the output impedance. However, the larger open-loop gain of the op amp makes the assumptions above much more accurate. (An op amp having an open-loop gain of 1,000,000 has an input current I_i much more insignificant than an op amp with an open-loop gain of 1000.)

If a phonograph pickup having an internal impedance of 50 ohms produces 0.001 volt across an open circuit and it is desired to amplify that output to 1.0 volt, a 50-ohm resistance would be placed between the pickup and an op amp −input, as in Fig. 10-7. Half of the pickup's open-circuit voltage will then be "lost" across its internal impedance, leaving the signal voltage V_s to be 0.0005

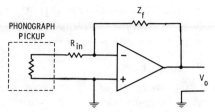

Fig. 10-7. Op amp used for amplification of phono pickup.

volt. The required amplification, or gain, is 2000 (1.0 ÷ 0.0005). With Z_{in} set at 50 ohms, the feedback impedance Z_f must have a value of 100,000 ohms to create a Z_f/Z_i ratio of 2000.

Negative feedback (Fig. 10-6) is also used in correcting deviations in frequency response. Many phonograph pickups do not react equally at the higher audio frequencies. Typically a pickup will reproduce audio waves (the electrical equivalent of sound waves)

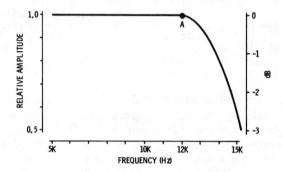

Fig. 10-8. Frequency response of phonograph pickup.

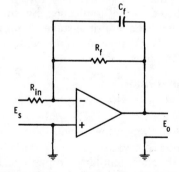

Fig. 10-9. Audio op amp with compensating feedback circuit.

having frequencies above 12,000 Hz at lower voltage levels. Fig. 10-8 shows the pickup output being fairly constant up to about 12,000 Hz at point A. Above this frequency the output decreases to 0.707 of its normal voltage level (a 3-dB loss) at 15,000 Hz. Correction of this loss can be attained by adding a reactance to the

107

feedback impedance (Fig. 10-9). This feedback reactance is shown as a capacitance. Frequency response can be made variable by using a variable resistor for R_f.

RF Amplifiers

The wide frequency range of op amps make them equally usable as radio-frequency (rf) amplifiers. Rf amplifiers differ from audio amplifiers only in the form of the input and feedback impedances. Commonly these feedback impedances will be series or parallel resonant circuits like those shown by Fig. 10-10. Designed to have impedance values in the proper ratio equaling the desired closed-

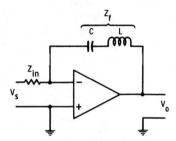

Fig. 10-10. An rf amplifier using an op amp.

loop gain, it may be necessary to also use resistances in series with these resonant circuits. Actually with the trend toward using as few inductors as possible, the circuit of Fig. 10-6 will also function for RF amplification. However, it must be recognized that Fig. 10-6 circuit is not selective but in many situations only one frequency is presented to the amplifier input. For example, converters of radio and television receivers develop a signal at the fixed intermediate frequency (if) for amplification—other frequencies are reduced to insignificance by tuned transformers of the converter. Thus, the multiple-stage if amplifiers of older tube and transistor receivers are now reduced to a single op amp with few discrete components.

Integrator

Among the initial uses of op amps was that of integration or summation. Care must be taken not to confuse the terms integration, integrator circuit, and integrated circuit. All have a common derivative meaning sum or total but each has a different reference. Similarly, in previous and following paragraphs the terms differentiator, differentiating, differentiating amplifier, differential amplifier, etc., require care in avoiding confusion. Regrettably the mathematics of integration is far beyond this book but it suffices to say that it is often necessary to determine the sum of values of varying factors. For example, in Chapter 2, Table 2-1, the sum and the mean of current squared values was determined as a rough form of integration.

The integrator circuit (Fig. 10-11) utilizes the same basic negative-feedback circuit as the audio and rf amplifiers. Capacitor

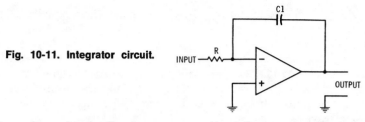

Fig. 10-11. Integrator circuit.

C1 used as the feedback component impedes current flow to a greater degree as its charge increases. In turn, the gain of the inverting amplifier increases as capacitor C1 acquires a charge. With a square-wave input, the constant voltage of the positive input pulse charges capacitor C1 at a steady rate permitting the output to increase negatively at a steady rate, from point A to point B in Fig. 10-12. Between points B and C, the negative input pulse

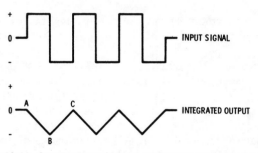

Fig. 10-12. Square-wave input and its integrated output.

discharges C1 and the output falls to zero. The resulting output—a triangular shaped wave—represents the negative integral (sum) of the input signal with respect to time. The output wave can be thought of as showing the instantaneous total of electrons that has entered the input—that total growing steadily from zero to a maximum, etc.

With a sine-wave input (Fig. 10-13) the integrator circuit is a bit more difficult to understand. Assuming the charging of C1 begins as the input signal is at its positive maximum, point A, the output is zero at that point. As the output is inverted, the charge of C1 controlling the gain is similarly inverted so that the output swings to a maximum negative value as the input signal becomes zero at point B. Following this through the complete cycle, the output

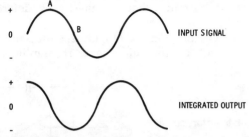

INPUT SIGNAL

INTEGRATED OUTPUT

Fig. 10-13. Sine-wave input and its integrated output.

of an integrator circuit is a cosine wave when its input signal is a sine wave.

Differentiator

Fig. 10-14 shows the differentiator circuit with a capacitor as the input impedance and a resistor for a feedback component. Initially, as the square-wave signal is applied, the capacitor offers little impedance and there is a surge of output voltage. A portion of that voltage surge is fed back through the resistor adding opposition

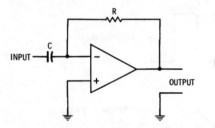

Fig. 10-14. Differentiator circuit.

to the capacitor impedance. Together the feedback and increase in input impedance quickly cuts off the output. At the end of the positive input pulse, capacitor C1 rapidly discharges and charges in the opposite direction to develop another brief surge of output

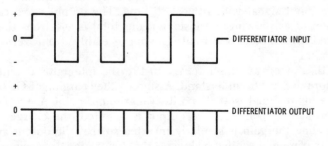

DIFFERENTIATOR INPUT

DIFFERENTIATOR OUTPUT

Fig. 10-15. Square-wave input and the differentiated output.

110

voltage. Fig. 10-15 shows the relation of these brief surges of output voltage—"blips"—and the square-wave input signal.

Noninverting Amplifier

In the circuits discussed above, the basic circuit has been that of an inverting amplifier with the input signal applied to the −input of the op amp. As might be expected, the noninverting amplifier has its signal applied to the +input of the op amp. The output of the noninverting amplifier is in phase with its input. However, this in-phase output does not provide for the usual feedback. Rather than feedback to the +input, the feedback is to the −input. The feedback signal then reacts with the input signal within the differential amplifier section(s) of the op amp. Fig. 10-16 shows a simplified noninverting amplifier. The input voltage V_s is developed

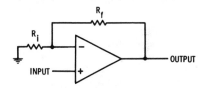

Fig. 10-16. Simplified noninverting amplifier.

across R_i, while the feedback voltage develops across the R_f and R_i series combination. Voltage gain of this noninverting amplifier is the sum of $R_f + R_i$ divided by R_i.

$$\text{Voltage gain, } V_o/V_s = (R_f + R_i)/R_i$$

Oscillators

The simplest type of oscillator using op amps has positive feedback between the output and the +input. Fig. 10-17 shows an oscillator using the inductive coupling between inductances L1 and L2 in the feedback circuit. Inductance L1 and capacitor C1 form a parallel resonant circuit determining the frequency of oscillation. In many actual circuits, there is also feedback to the −input as a

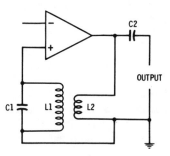

Fig. 10-17. IC oscillator using inductive coupling.

matter of stabilizing operation and reducing distortion. A number of op amp oscillator circuits are shown in Chapter 11.

ELECTRICAL CHARACTERISTICS

The following is a list of op amp electrical characteristics and their definitions.

Input offset voltage—Value of dc input-signal voltage at the differential input required to produce a zero output-signal voltage.

Input offset current—Difference between the two differential input bias currents when the output is zero.

Input bias current—Current supplied to each input of differential stage for proper biasing.

Device dissipation—Power dissipated by IC in form of heat—power lost.

Common-mode input voltage—Voltage common to the two inputs of a differential amplifier.

Common-mode rejection ratio—Ratio of open-loop gain to common-mode gain.

Open-loop voltage gain—The gain of an op amp operated without feedback.

Slew rate—The maximum rate of change of an op amp output with respect to time (of importance when working with square waves or very short-duration pulses).

Bandwidth (frequency range)—The band of frequencies within which an op amp functions as intended.

REVIEW QUESTIONS

1. What is an operational amplifier?
2. Were op amps ever built with electron tubes or discrete transistors? Name two reasons, in addition to size, for using ICs in op amps.
3. Describe common-mode rejection. Name two uses for common-mode rejection.
4. What purpose does the differential amplifier perform? Does the differential amplifier produce an output signal when common-mode signals are applied to both the +input and the −input?
5. Describe the outputs of a differential amplifier when the −input is grounded and a small ac signal is applied to the +input.
6. What is an inverter circuit? What is a Darlington pair?
7. What was the initial intended purpose of an op amp?
8. What is the open-loop gain of an op amp? The closed-loop gain, when negative feedback is used?

9. If a magnetic tape playback head has an internal impedance of 150 ohms and produces a maximum of 0.5 volt across an open circuit, what values should be used for input and feedback resistors to develop an output of 5 volts?
10. How does an rf op-amp amplifier differ from an audio op-amp amplifier?
11. What is the basic function of the integrator circuit? When a square wave is applied to the input of an integrator circuit, what is its output? A sine-wave input?
12. What is a differential? What is the output wave of a differentiator when a square wave is applied?
13. What is the voltage gain of a noninverting amplifier having a feedback resistor of 400,000 ohms and an input resistor of 800 ohms?

CHAPTER 11

Radio Frequency Production and Radiation

Long before the electronic era began, James Clerk Maxwell, a Scot (1831-1879), advanced the electromagnetic wave theory. Maxwell thought that high-frequency electromagnetic waves could be radiated through space and detected at some distant point. Heinrich Hertz, a German physicist (1857-1894), produced high-frequency waves using an inductance coil and a spark gap to prove Maxwell's theory of radiation. On June 2, 1896 Marconi was issued the first patent on wireless telegraphy based upon those early experiments of Hertz. Within a short span of years nearly every seagoing vessel was required to be equipped with a wireless telegraph spark-gap transmitter. Many of the spark-gap transmitters were retained in service aboard ships as late as the 1940s. As the first generator of high-frequency electromagnetic waves, the spark-gap oscillator is historically significant.

THE OSCILLATOR

Alexander Graham Bell invented the telephone in 1876 by sending the human voice over wires. After the wireless telegraph managed to transmit dots and dashes the question was whether or not speech could also be transmitted over it. However, it was the "telephone howler" that gave deForest and others their early clues toward the development of an electron tube oscillator to produce high-frequency electromagnetic waves.

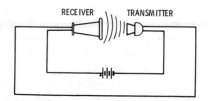

Fig. 11-1. A telephone receiver and transmitter arranged to produce an audible feedback.

The telephone howler is shown in Fig. 11-1. By locating the receiver so that a portion of its output is fed back (returned) into the transmitter, a howl is produced by the self-exciting action. The telephone howler illustrates the principle of an energy, or a force, being fed back to its point of origin to produce oscillations.

Fig. 11-2. Block diagram showing the operating principle of the oscillator circuit.

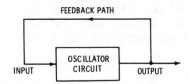

The principle of the oscillator is shown in Fig. 11-2. Here, a portion of the amplifier's output is fed back to its input to sustain oscillations. Since the only moving elements of the transistor amplifier circuit are its current carriers, there is almost no limit to the rate of oscillations that can be produced.

Transistor Oscillator

The oscillator circuit is primarily a radio frequency power amplifier arranged so that a portion of its output becomes its input signal. A simplified form of a transistor rf amplifier is shown by Fig. 11-3. Generator G supplies an alternating voltage between the base and

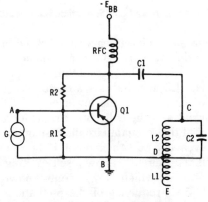

Fig. 11-3. A basic amplifier circuit.

115

the emitter (points A and B) of transistor Q1. When point A becomes negative with respect to point B the collector current increases and point C becomes negative with respect to point D. Since points B and D are electrically the same, voltage variations of points A and C are always opposite in polarity, or 180° out of phase. The current through L2, between C and D, produces magnetic lines of force that induce a voltage into the lower section of the winding (L1) between D and E. Just as point C is negative with respect to point D, point D is negative with respect to E. Stated in another way, E is positive with respect to point D when point A (the base) is positive with respect to B. Accordingly, points A and E are always of the same polarity—in phase. With a proper number of inductor turns between points D and E, the voltage induced between these points will equal that of generator G and can supplement that voltage, as shown by Fig. 11-4. The base-emitter junction is forward-biased allowing current to flow

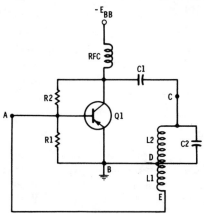

Fig. 11-4. A self-exciting amplifier circuit.

in the collector circuit to charge capacitor C2. Discharge of capacitor C2 through inductance L2 starts a resonating current within the L2-C2 combination. The resonating or alternating current through inductance L2 also induces an alternating voltage within inductance L1. When point E, the lower end of L1 becomes negative, the base current and the collector current increase to maintain the flywheel effect of the resonant circuit. Base and collector currents are reduced or even cut off when point E becomes positive with respect to point B. This circuit is now a self-exciting amplifier, or an oscillator—a variation of which is commonly known as the Hartley oscillator (Fig. 11-5). Frequency of the oscillations can be altered by changing the value of either the inductance or the capacitance.

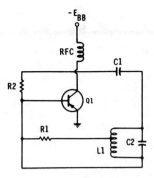

Fig. 11-5. Hartley oscillator circuit.

MOSFET Oscillator

Except for the omission of the biasing resistors R1 and R2, the MOSFET oscillator circuit in Fig. 11-6 is much the same as that in Fig. 11-4. Since the MOSFET is of the depletion type, a current begins to flow instantly through the channel—from the source to the drain and into the L2-C2 resonant circuit. In turn a voltage is induced into inductance L1 to alternately vary the potential on

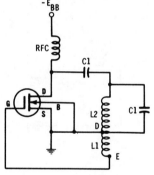

Fig. 11-6. MOSFET oscillator circuit.

the gate. The drain current then has an alternating component that will be opposed by the reactance of the rfc (radio frequency choke) inductance and readily passed by the coupling capacitor C1. Therefore, the alternating component of the drain current goes to the L2-C2 resonant circuit and its frequency is controlled by values of L2 and C2.

Other Oscillator Types

The only electron-tube–type oscillator of importance today is the *magnetron* oscillator. Electrons from the magnetron tube's cathode are set into a spiral motion by an external magnetic field. In turn, electrons enter resonant cavities of the anode causing the spiral to have an oscillating characteristic. The one resonant cavity (Fig. 11-7) having an outlet passes electrons in that oscillating manner.

Fig. 11-7. Magnetron tube.

Dimensions of the resonant cavities govern the frequency—300 MHz to 30 GHz—of the comparatively large power produced. It is this large production of power that makes the magnetron oscillator extremely useful for microwave relay of television signals, and for microwave ovens.

As might be suspected, the IC Op-Amp has become an important oscillator component. By having both an inverting and a noninverting input, it becomes fairly simple to design a feedback circuit to sustain oscillations. An op-amp oscillator circuit commonly feeds a portion of the output back into its noninverting input (+) through some form of coupled-resonant circuit (Fig. 11-8A), series-

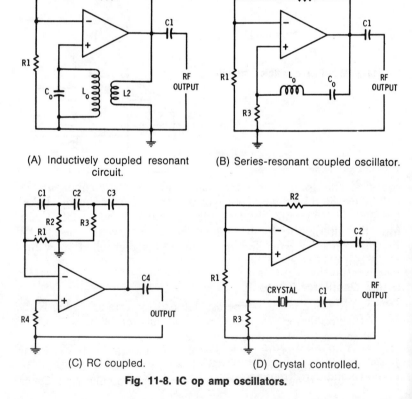

(A) Inductively coupled resonant circuit.

(B) Series-resonant coupled oscillator.

(C) RC coupled.

(D) Crystal controlled.

Fig. 11-8. IC op amp oscillators.

resonant circuit (Fig. 11-8B), RC filter (Fig. 11-8C) or crystal-controlled circuit (Fig. 11-8D). Such feedback devices select the frequency of oscillations. While the oscillations are produced by the noninverting amplifier, voltage or current fed back to the inverting input by way of R1 and R2 in Figs. 11-8A, B and D (Fig. 11-8C is basically an audio frequency oscillator) improves the stability and reduces distortion.

Voltage-controlled oscillators are finding greater usage as we proceed into our programmed and computerized era. Simplified, the voltage-controlled oscillator (vco) uses a phase-shifter voltage to alter the oscillator's frequency. Fig. 11-9 shows a portion of the oscillator output fed into the emitter of transistor Q1, and the base current determined by an applied voltage E_s. In turn, the amount

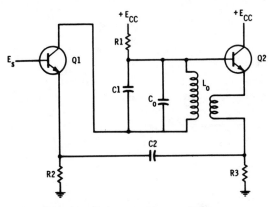

Fig. 11-9. Voltage-controlled oscillator.

of rf current going to the collector-emitter circuit varies with E_s. Passing through R1 and C1, the rf current produces a voltage across C1 that is 90° out-of-phase with that across L_o. In effect, this out-of-phase voltage adds reactance to the L_o-C_o combination and changes its resonant frequency in relation to the amplitude of input voltage E_s. Since voltage E_s can be derived from a great many factors, the voltage-controlled oscillator has many uses. A resistor varying with temperature provides the control voltage E_s of a vco, while a digital counter translates the vco output to visual numerals showing the actual temperature in either Fahrenheit or centigrade. And, of course, voltages can be measured by using the unknown voltage directly or indirectly as the control voltage of a vco providing pulses to a digital counter.

Closely related with the voltage-controlled oscillator is the *phase-lock-loop* (PLL). Output of the oscillator is compared with a known standard frequency and any difference is used as the con-

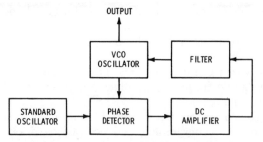

Fig. 11-10. Block diagram of a phase-locked-loop circuit.

trol voltage (Fig. 11-10). Assuring stability, the PLL circuit is available in the integrated circuit form.

RADIATION

After watching the traffic on our streets and highways it becomes apparent that vehicles travel more or less in bunches. Traffic lights or slow moving vehicles tend to hold back the normal flow of traffic until a group has formed and moves on down the highway. Between these groups are likely to be intervals of time when no vehicles will pass a particular point. If the density of traffic along a city street is plotted, as in Fig. 11-11, there appear to be waves of traf-

Fig. 11-11. Graphic representation of automobile traffic moving in one direction on a typical city street.

fic. Assigning positive or negative notation to the direction of traffic flow, these waves of traffic may acquire the appearance of those plotted in Fig. 11-12.

Fig. 11-12. Graphic representation of automobile traffic moving in both directions on a typical city street.

Radio-Wave Characteristics

Much like this vehicular traffic, the current caused by an alternating emf travels through a conductor in waves. Suppose an emf that alternates at 100 Hz produces an alternating current of 50 amperes maximum. During the first one-tenth of the cycle, 0.001

120

second, current flow is about 15 amperes—corresponding to a movement of 9,420,000,000,000,000 electrons. If we assume that this group of electrons moves at the speed of light (186,000 miles per second) it will travel 186 miles during the first one-tenth of the cycle. Joined by other electron groups varying in size according to the trigonometric sine function, each group will move 186 miles through the conductor during each succeeding period of 0.001 second (Fig. 11-13). These groups of electrons move through a conductor much like a wave and are referred to as *waves of current*.

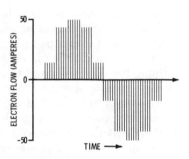

Fig. 11-13. Waves of current flow through a conductor at 1/10 second increments.

Wavelength

If the wave of current having a frequency of 100 Hz moves 186 miles in one-tenth of the time for one cycle, the distance traveled during a complete cycle will be 1860 miles or 3,000,000 meters. Thus, a frequency of 100 Hz is said to have a *wavelength* of 1860 miles, or 3,000,000 meters. Since a wave having a frequency of 200 Hz completes its cycle in half the time of a 100-Hz wave, the wavelength of a 200-Hz frequency will be 1,500,000 meters. And 300 meters is the wavelength of a wave having a frequency of 1,000,000 Hz. Wavelength then varies inversely with its frequency according to the equation:

$$\text{Wavelength} = \frac{300,000,000}{\text{frequency}} \text{ meters}$$

or,

$$\frac{186,000}{\text{frequency}} \text{ miles}$$

In the above discussion it is assumed that a wave of current flow travels at the speed of light. In actual practice, however, it will be somewhat less because the normal resistance and any reactance that might be associated with the conductor will restrict this speed. The characteristic of the conductor thus determines the *true wavelength* while the standard wavelength, as given by the equation, is determined at the speed of light.

Basically, wavelengths are classified in four groups—long wave, medium wave, short wave, and microwave. Long waves are those having lengths greater than 500 meters and frequencies below 600,000 Hz. Between 188 and 500 meters are the medium waves with frequencies from 600,000 to 1,600,000 Hz, the Standard Broadcast band. Wavelengths shorter than 188 meters are known as short waves while microwave is the term given to those wavelengths of less than one meter.

Standing Waves

A current wave is somewhat analogous to a wave of water. If a disturbance is caused at one end of a long canal, waves travel uninterrupted along the full length of the canal until they dissipate. If, however, a dam is placed at some point in the canal, the waves striking the dam will react to cause a second wave which will move in the opposite direction. Very similarly a current wave that suddenly comes to the end of a conductor will be blocked. But as you know, any current flow produces a magnetic field and any time that current flow is changed or stopped, the magnetic field is correspondingly altered. However, a changing magnetic field produces an emf to oppose that change. This self-induced emf will then cause a current wave to move in the opposite direction—a *reflected wave*. Fig. 11-14 shows the initial wave moving to the right while the wave moving to the left is a reflected wave. By adding the instantaneous values of the initial and the reflected waves the resultant wave is found. This wave (indicated by the dotted line) remains in the same position and therefore is considered a *standing wave*. The standing wave will vary in magnitude. Naturally, if the conductor is terminated in a load or an impedance that absorbs the initial wave, there can be no reflected wave and hence no standing wave.

Although the standing wave is the resultant of two waves of current, it does have a corresponding magnetic field that varies in

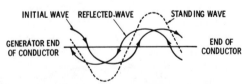

Fig. 11-14. A standing wave produced by feeding alternating current through a conductor with no terminating "load."

intensity. When a current at frequencies above 200,000 Hz is fed to a conductor, it causes a form of electromagnetic energy known as radio waves to be radiated from it. These waves are comprised of electric and magnetic fields of energy which move through space with the speed of light. The characteristics of these waves is such that current will be induced in any conductor they contact. The conductor that radiates these waves and the one into which their forces are induced are termed the *transmitting* and *receiving antennas*, respectively.

Propagation

The wavelengths of frequencies below 200,000 Hz are so long that it is difficult to obtain radiation into space. The atmosphere normally dampens all radiated waves to some extent, but those which are very long diminish quite rapidly. Frequencies between 200,000 and 500,000 Hz produce waves having a tendency to disperse—in contrast to the extremely short waves of light that travel in a straight line. Long waves tend to curve around obstructions and even follow the curvature of the earth to some degree.

Thus, long-wave radiation is generally used for those radio applications that must be dependable (transoceanic telephone, radionavigation of ships and aircraft, etc.). Although one normally thinks of the standard broadcast station as being dependable, it is limited to about 30 miles for its primary service area. This area can be extended by increasing the amount of radiated power; however, the communicating ranges does not increase in direct proportion to this increase. Waves which travel upward from the earth's surface are known as *sky waves*. Under the proper conditions, these waves may be refracted (bent) back to the earth many miles from their point of origin.

Reception of this refracted sky wave is unpredictable. However, sky wave, or "skip" reception as it is sometimes called, at frequencies up to 30,000,000 Hz is generally good enough for extensive commercial communication. At frequencies above 30,000,000 Hz sky waves are refracted very little. Thus, sky waves at the higher frequencies do not normally return to earth. Our television stations, operating at frequencies above 50,000,000 Hz are generally limited to a service area having a radius of less than 50 miles.

While the higher frequencies have the advantage of limited range, the fact that these waves do not bend readily makes them desirable for point to point communications. By concentrating these waves into a beam, much like that of a flashlight, they can be directed to a particular area. These higher frequencies have also been found advantageous for local communications (police and fire departments, taxis, etc.) where distance is not too important.

In the microwave range, above 300,000,000 Hz (300 megahertz), radio waves can be concentrated to such an extent that on striking a solid object, such as an airplane, the beam reflects back to a receiver to indicate the object's presence and location. This is the

123

operating principle of *Radar*. *Sonar* uses the same principles but with very-low frequency waves (just above the audible range) acting in the medium of water.

To further illustrate the propagation characteristics of the various frequency bands, Table 11-1 lists the primary uses and advantages or disadvantages.

Antenna-Radiation Characteristics

Suppose now that we apply an alternating emf having a frequency of 300 megahertz to the lower end of a vertically mounted rod conductor 22.5 centimeters long. The length of this rod then equals 0.225 wavelength at this frequency. The magnetic lines of force in the waves radiated from this rod will be at right angles with the rod and radiation from the ends will be negligible. With the rod mounted vertically, as in Fig. 11-15, the intensity of the radi-

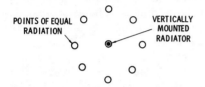

Fig. 11-15. Radiation characteristics of a simple vertically mounted radiating element.

ations at all points equidistant from and on a horizontal plane about the rod will be equal. Thus, a vertically mounted rod is said to produce a nondirectional radiation pattern. Since radiation is not produced from the ends of the rod, a horizontal mounting (Fig. 11-16) results in a bidirectional radiation pattern.

Ground Conductivity

The earth, or ground, is in reality the return path of the radiated wave. Thus, the conductivity of the earth about the transmitting and the receiving antenna is a factor in the transmission and reception of radio signals. In areas having sandy soil, such as Florida,

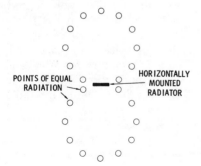

Fig. 11-16. Radiation characteristics of a simple horizontally mounted radiating element.

Table 11-1. Uses and Characteristics of Various Frequency Bands

Band	Primary Uses	Characteristics
Very low frequency (vlf) 10 to 30 kHz	Long-range communication	Reliable. Ground-wave attenunation small. Sky-wave reflection good. Antennas large and expensive.
Low frequency (lf) 30 to 300 kHz	Shorter-range communications	Some ground attenuation. Less ionosphere reflection of skywaves. Good nighttime transmission.
Medium frequency (mf) 300 to 3000 kHz	Aircraft and weather communication Standard am broadcasting (550 to 1600 kHz)	Dependable for distances up to 100 miles. Some sky-wave reflection.
High frequency 3 to 30 MHz	Class D Citizens Band—CB (26.965 to 27.405 MHz) Amateur radio	Some sky-wave and ionospheric reflection. Ground-wave coverage of 25 mile (40-km) possible.
Very high (vhf) frequency 30 to 300 MHz	Commercial fm and vhf television channels (fm, 88 to 108 MHz) (tv, 54 to 72 MHz, 76 to 88 MHz, 174 to 216 MHz) Aviation services (108 to 136 MHz) Marine, public safety, industrial and business Amateur radio (150 MHz)	Basically "line-of-sight" propagation. Some atmospheric refraction—not dependable.
Ultrahigh frequency (uhf) 300 to 3000 MHz	Television (470 to 807 MHz) Microwave cooking Microwave relaying of telephone conversations, television signals, etc.	Primarily "line-of-sight" propagation but somewhat extended by atmospheric refraction.
Superhigh frequency (shf) 3000 to 30,000 MHz 3 to 30 GHz	Basically experimental	Lasers, masers and light beams fall within this range and are just now being utilized. While today's telephone conversations may travel via such beams, we are not aware of such things.

it can be noted during a rain shower or immediately afterwards that reception is excellent; however, the sandy soil soon loses that conductivity and reception becomes poor. Therefore, the transmitting antenna is often erected in swampy areas or near the edge of a body of water. Sometimes the antenna tower may even be mounted in the water. Many transmitting stations use an elaborate system of wires placed underground to further ensure conductivity. Modern home receivers make use of the grounded side of the power line which probably is as good or better than the average ground system the owner might install.

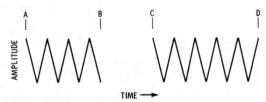

Fig. 11-17. Interrupted radiating waveform.

The theory of radiation presented here has been primarily based on that advanced by the German scientist Heinrich Hertz (1857-1894) following that given by James C. Maxwell. However, the Quantum theory developed by two Germans named Max Plank and Albert Einstein cannot be omitted. The two theories differ only in the form taken by the radiations once they are emitted into space. According to the Quantum theory all energy exists in bundles, called quanta, which are radiated from the antenna. These quanta do not stimulate any of the five senses and are therefore highly abstract. Work with atomic energy and its radiations has confirmed much of the Quantum theory and it seems that the two theories may eventually be found to coincide.

MODULATION OF RADIATED WAVES

Before a radiated wave can be of importance it must contain or carry intelligence or information. Just as the automobile and the airplane would be useless if they did not carry passengers or cargo, the radiated wave is of little use without some form of information "loaded" or *modulated* onto the wave. This modulating takes on

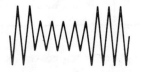

Fig. 11-18. Wave with reduced-amplitude form of interruption.

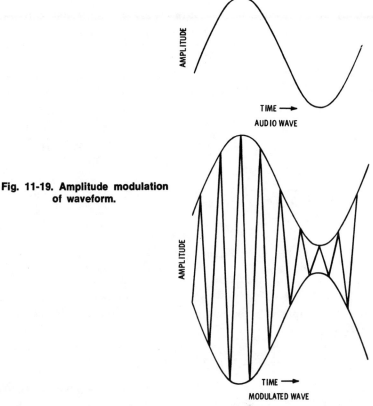

Fig. 11-19. Amplitude modulation of waveform.

many different forms. Wireless telegraphy interrupted the radiated wave to provide an early form of modulation. Fig. 11-17 shows this interrupted wave with a long period of radiated wave representing a dash and a shorter period representing a dot, separated by a period of no radiation. Another form of this interrupted wave, Fig. 11-18, reduces the wave's amplitude rather than cutting it off entirely. The wave of Fig. 11-18 can be considered to be a form of *amplitude modulation.* However, true amplitude modulation is shown by Fig. 11-19, with both increases and decreases in the waves amplitude. As shown by Fig. 11-19 the amplitude is varied according to an audio or sound wave and is typical of the modulated waves radiated by our standard broadcast stations. The wave carrying the video or picture of a television broadcast station is also amplitude modulation. Frequency modulation (fm) broadcast stations modulate their radiated waves by varying the frequency. Fig. 11-20 shows a wave of constant amplitude starting at one frequency, dropping to a lower frequency, returning to the original frequency, and then increasing to a higher frequency. This could readily be the modulating result of an alternating signal as it starts at zero, rises to a maximum, drops back to zero, goes to an opposite maximum, etc. Fm broadcast stations modulate or vary their frequency by as much as 0.075 MHz with passage of the greatest sound volume. That is, the maximum volume of the symphony orchestra might cause the frequency to vary or modulate 0.075 MHz, while

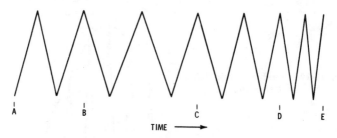

Fig. 11-20. Frequency-modulated waveform.

the modulation under similar circumstances might be only a few kilohertz for the notes from a piccolo. Frequency modulation differs from amplitude modulation since it has no variation in amplitude.

REVIEW QUESTIONS

1. Who first developed the theory of radiating electromagnetic waves?
2. In what year did Marconi receive his patent on wireless telegraphy?
3. Describe the telephone howler.
4. What is the basic principle of an oscillator? What is meant by fed-back current or voltage? Name two conditions that must be met by this fed-back current in order to support oscillations.
5. What type circuit is used for the oscillator?
6. Name one basic type of oscillator circuit.
7. How can the frequency of an oscillator be changed?
8. Does the MOSFET oscillator differ greatly from the bipolar transistor circuit?
9. What is wavelength?
10. What is the approximate speed at which radio waves travel through space?
11. What is a reflected wave?
12. Are radiocommunications more reliable or less reliable at frequencies *above* 30,000,000 Hz (30 megahertz)? Explain.

13. At what position, with respect to the earth, would a simple radiation element have to be placed to provide a nondirectional radiation pattern?
14. What is modulation? What is the difference between amplitude modulation and frequency modulation?

CHAPTER 12

Digital Circuits

A quick look at the series of lines on our cereal box is the only indication needed to realize how important computers are in our lives. And computers are digital circuits. However, digital circuits are presently being used or being introduced into so many fields that we need not consider the complete computer. In fact, it would be very impractical to study the computer.

BINARY ARITHMETIC

Having been taught to count by the decimal system—1, 2, 3, 4, 5, 6, 7, 8, 9, 10, etc.—it is a bit difficult to conceive of a system using only two digits or symbols. First, recognize that 1, 2, 3, etc., are only symbols or digits that represent quantities. Our common system has ten digits—0, 1, 2, 3, 4, 5, 6, 7, 8, and 9—and is referred to as the decimal, or decinary, system. The number 493 is said to have three places or three bits and its least significant bit is at the far right. The most significant bit is at its far left.

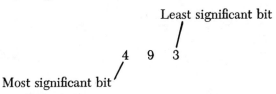

The least significant bit indicates the quantity of units included in the complete number 493. The next to least significant bit indicates the quantity of tens—in this case, there are 9 tens. Other bits represent the quantity of hundreds, thousands, ten-thousands, hundred-thousands, millions, etc.

$$\begin{array}{r} 3 \text{ units} = 3 \\ 9 \text{ tens} = 90 \\ 4 \text{ hundreds} = 400 \\ \hline 493 \end{array}$$

Since digital circuits respond most readily to only two conditions, their number system has only two digits—0 and 1—and is known as the binary system. The binary number 1001 has four places or four bits. Unlike the decimal number, the least significant bit is not always at the far right—it can be at the far right or far left. For our study, the least significant bit will be at the far right.

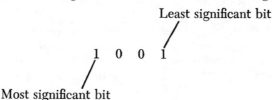

Like the decimal system, the least significant bit indicates the quantity of units in the complete number. With only two digits used in the binary number, and 0 indicating no units, the greatest quantity of units that can be indicated in the least significant bit is one. So the next bit must indicate the quantity of twos. In turn, the third bit indicates the quantity of fours and the fourth bit indicates the quantity of eights. (Binary numbers may have additional bits indicating the quantities of 16s, 32s, 64s, and higher powers of twos.)

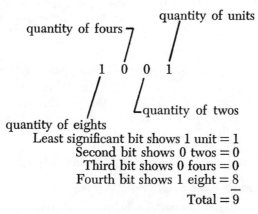

$$\begin{array}{r} \text{Least significant bit shows 1 unit} = 1 \\ \text{Second bit shows 0 twos} = 0 \\ \text{Third bit shows 0 fours} = 0 \\ \text{Fourth bit shows 1 eight} = 8 \\ \hline \text{Total} = 9 \end{array}$$

In other words, the binary number 1001 is equivalent to the decimal number 9.

binary number	decimal equivalent
0001	1
0010	2
0011	3
0100	4
0101	5
0110	6
0111	7
1000	8
1001	9
1010	10
1011	11
1100	12
1101	13
1110	14
1111	15

The two digits of the binary system corresponds with the two electrical states—on and off. Thus, the binary number can be represented by on and off conditions of current or voltage. A simple series circuit (Fig. 12-1) can be used to develop pulses of current

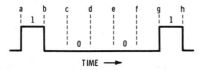

Fig. 12-1. Simple series circuit developing binary pulses.

in a manner representing the binary number. The pulses between times (a) and (b) and (g) and (h) of Fig. 12-2 indicate the digit 1, while the lack of pulses between (c) and (d) and between (e) and (f) shows digits 0 and 0. In this manner, the train of pulses

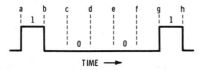

Fig. 12-2. Graphic representation of binary pulses produced by series circuit.

and lack of pulses represent the binary number 1001. The single lamp of the Fig. 12-1 circuit would blink to that same pattern.

Switches S1 and S4 are closed in the parallel arrangement of Fig. 12-3 to light lamps B1 and B4, and represent the digit 1 in the first and fourth bits. Bits represented by lamps B2 and B3 are not lit since switches S2 and S3 are open. In this way, the four lamps show the binary number 1001 continuously.

Binary Addition

Before any number system can be practical, it must be possible to perform the four basic mathematic operations—addition, subtraction, multiplication and division. Binary addition follows the same

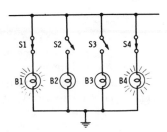

Fig. 12-3. Parallel representation of binary number.

basic rules as decimal addition. In the decimal system, the maximum quantity that can be held by each position or bit is represented by the digit 9. Adding a decimal 1 to a position holding a 9 digit empties that position and carries a 1 to the next position. That is, 1 added to the decimal number 49 empties its least significant bit and carries 1 to the next bit.

$$
\begin{array}{r}
4\ 9 \\
\text{plus}\quad 1 \\
\hline
1\quad \text{carry} \\
\hline
5\ 0\ \text{sum}
\end{array}
$$

A binary bit is filled by the quantity represented by the digit 1. When another unit is added to the binary number 1001, its first bit is emptied and a 1 carried to the second bit.

$$
\begin{array}{r}
1\ 0\ 0\ 1 \\
\text{plus}\quad 1 \\
\hline
1\quad \text{carry} \\
\hline
1\ 0\ 1\ 0\ \text{sum}
\end{array}
$$

The possible combinations of binary addition are indicated in the following table.

First number	Second number	Sum	Carry
0	0	0	0
0	1	1	0
1	0	1	0
1	1	0	1

In situations where the carry puts a third number into the column, add the first two and then use that resulting sum as the first number of another addition.

$$
\begin{array}{r}
1\ 0\ 0\ 1 \\
\text{plus}\ 0\ 0\ 1\ 1 \\
\hline
1\ 1\quad \text{carry} \\
\hline
1\ 1\ 0\ 0
\end{array}
$$

Binary Subtraction

It is possible to subtract one binary number from another in the same manner as decimal numbers are subtracted. However, such direct subtraction of binary numbers requires more actual reasoning than the electronic computer is capable of. Regardless of a computer's apparent capability, it is able to perform only as the programmer or operator directs—it has no ability to reason or think. Binary subtraction makes use of the complement of the subtrahend. In the problem of subtracting from 1001 the binary 0011, 1001 is the minuend and 0011 is the subtrahend. The complement of a binary number changes its 1 digits to 0s and its 0 digits to 1s—the complement of 0011 is 1100. Then by adding the minuend and the complement of the subtrahend the remainder is found.

$$
\begin{array}{ll}
1\ 0\ 0\ 1 & \text{(minuend)} \\
1\ 1\ 0\ 0 & \text{(complement of subtrahend)} \\
\hline
(1)\ 0\ 1\ 0\ 1 & \\
1 & \text{end-around carry} \\
\hline
+\ 0\ 1\ 1\ 0 & \text{remainder}
\end{array}
$$

The end-around carry is not a part of a football play but an adjustment for the forming of the complement (this is also similar to borrowing a 1 from the next position as we do in decimal subtraction). An end-around carry also indicates a positive difference while the absence of an end-around carry indicates a negative difference that requires another complementing step.

In actual computers all negatives are converted to the complement, with a 1 added. This eliminates the end-around carry and the machine always adds rather than subtracts.

Binary Multiplication

Multiplication of binary numbers follows the same basic method as decimal multiplication. A primary difference in the multiplication tables is that the binary system has only four possible combinations.

$$
\begin{aligned}
1 \times 1 &= 1 \\
1 \times 0 &= 0 \\
0 \times 0 &= 0 \\
0 \times 1 &= 0
\end{aligned}
$$

$$
\begin{array}{cccc}
\begin{array}{r}
1\ 0\ 0\ 1 \\
\times\ 0\ 0\ 0\ 1 \\
\hline
1\ 0\ 0\ 1
\end{array}
&
\begin{array}{r}
1\ 0\ 0\ 1 \\
\times\ 0\ 0\ 1\ 0 \\
\hline
0\ 0\ 0\ 0 \\
1\ 0\ 0\ 1 \\
\hline
1\ 0\ 0\ 1\ 0
\end{array}
&
\begin{array}{r}
1\ 0\ 0\ 1 \\
\times\ 0\ 1\ 0\ 0 \\
\hline
0\ 0\ 0\ 0 \\
0\ 0\ 0\ 0 \\
1\ 0\ 0\ 1 \\
\hline
1\ 0\ 0\ 1\ 0\ 0
\end{array}
&
\begin{array}{r}
1\ 0\ 0\ 1 \\
\times\ 1\ 0\ 0\ 0 \\
\hline
0\ 0\ 0\ 0 \\
0\ 0\ 0\ 0 \\
0\ 0\ 0\ 0 \\
1\ 0\ 0\ 1 \\
\hline
1\ 0\ 0\ 1\ 0\ 0\ 0
\end{array}
\end{array}
$$

The above examples indicate that binary multiplication is quite similar to the decimal situation of a multiplier having a number of 0s and 1s.

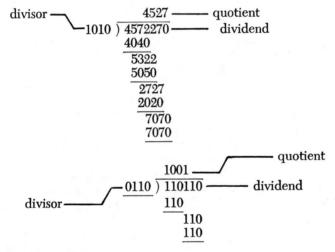

4527 (multiplicand) —— 1001 (binary equivalent of 9)
1010 (multiplier) —— 0110 (binary equivalent of 6)
―――――――――――――   ――――――――――――
0000   0000
4527   1001
0000   1001
4527   ――――――
―――――――――   110110 (binary equivalent of 54)
4572270 (product) —――/

Binary Division

Division is the opposite of multiplication and binary division is basically the same as decimal long division.

divisor ―――\   4527 ―――― quotient
          ――1010) 4572270 ―――― dividend
                     4040
                     ――――
                      5322
                      5050
                      ――――
                       2727
                       2020
                       ――――
                        7070
                        7070
                        ――――

                                         ―――― quotient
                         1001 ――――/
             ―0110) 110110 ――――― dividend
divisor ―――――/   110
                     ――――
                         110
                         110
                         ――――

Mechanics of Addition

The mechanical process of binary addition can be handled in either of two ways. Series addition is basically the same as counting on our fingers—one unit at a time. For such series binary addition, each bit or electrical pulse is fed into a circuit known as a *half adder*. That half adder consists of one OR gate, two AND gates, and an inverter as shown by Fig. 12-4A. Fig. 12-4B shows the circuit details of an AND gate—all inputs must be at the 1 state before the output goes to the 1 state. Circuit details of an OR gate are shown by Fig. 12-4C—a 1 state on any of its inputs puts a 1 state on the output. Then a pulse (a 1 state) applied to input A passes through the OR gate to the AND gate A2. But that pulse applied also to the A input of AND gate A1 does not develop a 1 state at A1 output.

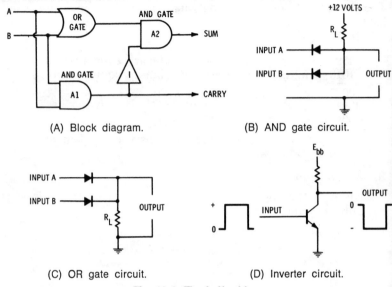

(A) Block diagram.

(B) AND gate circuit.

(C) OR gate circuit.

(D) Inverter circuit.

Fig. 12-4. The half adder.

Inverter I (Fig. 12-4D) does change the 0 state of A1 output to a 1 state. Thus, the OR gate and the inverter both feed pulses into AND gate A2 and it produces a 1 state to the sum. That sum state is then applied to input B (probably via a delay). A second pulse applied to input A passes to A2 and to inverter I putting 1 and 0 states on the two inputs of A2. With 0 and 1 inputs, A2 has a 0-state output. AND gate A1, with 1 and 1 inputs, feeds a pulse to the carry and input A of a second half adder. Fig. 12-5 shows the sum of half adder Ad1 going to an indicator lamp B1 (actually there would be a memory device between the half adder and the indicator lamp to retain the 1, or on-state condition). In succession, the first pulse would light B1, the second pulse turns B1 off and B2 on, the third pulse turns B1 on, the fourth pulse turns B1 and B2 off and B3 on

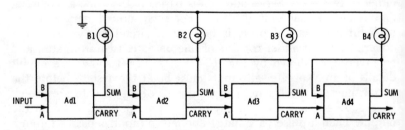

Fig. 12-5 Half adders acting in series to light indicator lamps in series binary addition.

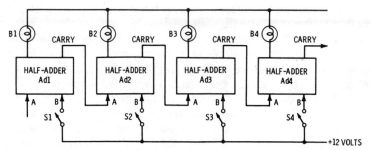

Fig. 12-6. Simplified parallel addition.

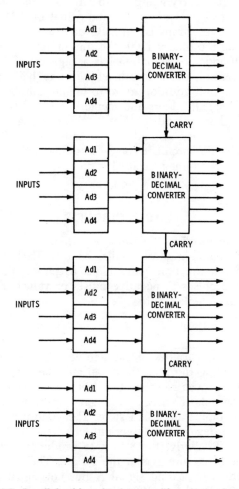

Fig. 12-7. Parallel adders for four-position decimal addition.

137

by way of half adder Ad3, etc. This is basic series addition of pulses applied to input A of half adder Ad1.

Parallel addition feeds a complete binary number—usually composed of four bits as 1001—to a corresponding number of half adders as indicated in Fig. 12-6. Switches S1, S2, S3, and S4 can be activated by a single push-button marked by the decimal numeral 9. In this manner, the decimal number 9 is converted to its binary equivalent of 1001. Depressing the 9 button lights lamps B1 and B4. If the 3 button is then pressed, half adder Ad1 acquires a sum of 0 and a carry of 1 going to Ad2. Ad2 with its 0 output, the 1 input from the push button, and the 1 carry from Ad1 develops a 0 output and a 1 carry going to Ad3. In turn, the carry from Ad2 puts the output of Ad3 to a 1 state and that of Ad4 remains in a 1 state. This leaves lamps B1 and B2 dark and lamps B3 and B4 lit, representing the binary 1100—the sum of 1001 + 0011 (9 + 3). This four half-adder combination has a capacity of 1111 (decimal 15) but the second, third, fourth, etc., sets of half adders, shown by Fig. 12-7, permit addition to go to infinite number of positions. Each set of four half adders and memory is wired to put a 1 carry to the next higher set and to clear as the sum reaches 1010 (10). In this way, the system also adds by decades—1, 10, 100, 1000, etc., and with conversion circuits to be discussed later the resulting sums appear in the accustomed decimal form.

BASIC CIRCUITS

As with other forms of electronic equipment, the binary circuits are usually a combination of smaller circuits. Each smaller circuit performs a given function—gates block or pass a signal, inverters change positive pulses to negative pulses or values, flip-flops store and shift binary numbers, etc.

Gates

There are many varieties of gates in addition to those shown in Fig. 12-4, including the NOR, the NAND and the exclusive OR gates. The NOR and the NAND gates are forms of the OR and AND gates having inverted outputs. That is, a NAND gate produces a 0-state output when all of its inputs are at a 1 state. This contrasts with the usual AND gate which develops a 1-state output when all of its inputs are at the 1 state. Similarly, the NOR gate produces a 0-state output when any one of its inputs is at the 1 state. The exclusive OR gate is very selective in passing signal A but not signal B or passing signal B but not signal A—neither signal passes if the other is present.

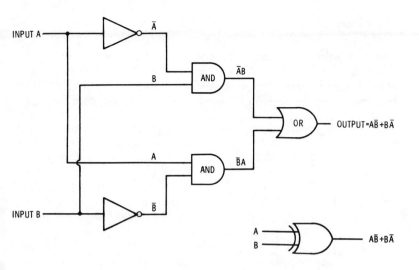

Fig. 12-8. Exclusive OR gate circuit.

The AND gate circuit in Fig. 12-4B uses diodes providing direct paths to any input terminal that is grounded or less than 12 volts positive. In turn, this pulls the output level down to less than +12 volt level—the 1-state level. However, at anytime when all of the inputs are at the 1-state level of 12 volts, the AND gate output will also be at that level. In this manner, the AND gate can be used to determine when conditions are correct for another action to occur. For example, in the digital watch or clock to be described later, as the binary-hour register reaches a sum of 1011 (13) an AND gate determines when binary positions 1, 2 and 4 are at the 1 state and resets that register to 0001.

Each input of an OR gate is in series with a diode and a common load resistor R_L. Whenever any of these inputs reach a 1-state level, the gate's output also goes to the 1 state (Fig. 12-4C). In contrast to the AND gate, the OR gate requires only one condition be satisfied before its output goes to the 1 state.

Fig. 12-8 shows the exclusive OR gate which is actually a combination of inverters, AND gates and OR gates. One AND gate has an A input and an inverted B input—meaning that if A and B are both at the 1 state, the inverted B is at 0 and the AND gate output is also at 0. Inputs of an inverted A and a B also develop a 0 at the second AND gate output. Thus, when A and B are both at 1, inputs of the OR gate are 0 and 0, and its output is 0. A 1-state output is produced by the exclusive OR gate only when inputs A and B are at different states—that is, when A is 1 and B is 0, or A is 0 and B is 1.

Flip-Flops

The most widely used element of any digital system is the flip-flop. A flip-flop is basically any electrical or mechanical device that is stable in either of two conditions—on or off, up or down, hot or cold. A playground seesaw, being worked properly, is a flip-flop, but the passenger elevator stopping at the many floors is not a flip-flop because it has more than two stable conditions. A typical electronic flip-flop uses two inverters as indicated by Fig. 12-9.

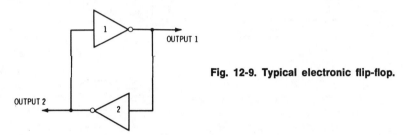

Fig. 12-9. Typical electronic flip-flop.

With the output of inverter 2 feeding back into inverter 1, it might be expected that the circuit would oscillate. However, this feedback causes inverter 1 to saturate and inverter 2 to cut off. Under these conditions the action soon becomes stable—the output of inverter 2 quickly goes to a 1 state, while that of inverter 1 attains a 0 state. In reality, the collector voltage of inverter 1 will not be zero but usually less than 1.5 volts—too low to be a 1 state.

Fig. 12-10 shows the flip-flop circuit in greater detail with a common input. Injection of a positive pulse—representing a binary

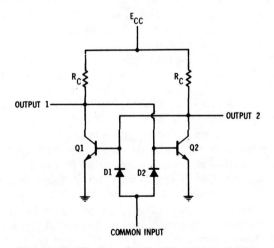

Fig. 12-10. Flip-flop circuit with common input.

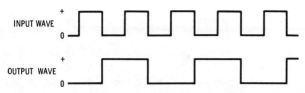

Fig. 12-11. Relations between flip-flop input and output.

1—into this common input clamps the base of each transistor to a high level (usually +12 volts) by way of the diodes D1 and D2. Inverter 2 then begins to conduct and the 1 and 0 states reverse. That is, the output of inverter 1 acquires a 1 state while inverter 2 goes to a 0 state. Each succeeding positive pulse alternates these output states as indicated by the graphic representation of Fig. 12-11. Note that two input pulses are required to put the flip-flop through a complete on/off—set/reset cycle. However, the flip-flop will remain at one state until another pulse is injected.

The characteristic of the flip-flop requiring two input pulses to produce one output pulse makes it usable as a frequency divider. If a flip-flop has an input of 2,097,152 pulses per second, its output rate will be 1,048,576 pulses per second. A second flip-flop in series will divide that rate by 2 to give 524,288 pulses per second (pps). In succession 21 flip-flops will each divide the rate by 2, until the 21st flip-flop has an output of 1 pps. There is also a system of feedback providing for frequency division by other than multiples of 2. For example, the two flip-flops shown in Fig. 12-12 have a natural binary ratio of N/4—meaning that the input frequency is divided

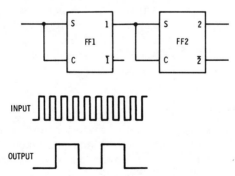

Fig. 12-12. Flip-flops with natural binary radio of N/4.

by 4. A feedback from the inverted output of flip-flop FF2 to a direct-set terminal of FF1 sets both flip-flops to the 1 state after the second pulse. That is, the first pulse sets FF1, the second pulse clears FF1 and sets FF2 but the feedback also re-sets FF1 (do not

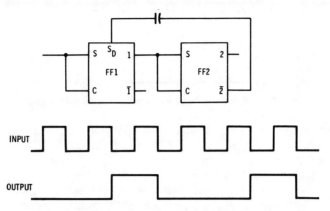

Fig. 12-13. Flip-flop with negative feedback producing a binary ratio of N/3.

confuse re-set with "reset" which is often used to mean "the 0 state"). Then the third pulse, rather than the fourth, clears FF1 which in turn clears FF2. In this manner, the combination of flip-flops in Fig. 12-13 has a binary ratio of N/3. In longer flip-flop series the feedback can be to any desired stage to provide a frequency division equaling the natural binary ratio minus those stages to which the feedback is applied. With a natural binary ratio of N/128, a series of seven flip-flops will have a binary ratio of N/107 if feedback is made to the fifth, third, and first flip-flops ($128 - 16 - 4 - 1$).

Pulses from a clock—regularly spaced with regard to time—fed to the inputs of a series of flip-flops shifts the data. For example, the binary number 1001 puts the outputs of FF1 and FF4 at the 1 state and FF2 and FF3 at the 0 state. A clock pulse fed to all of the flip-flops shifts each bit to the next position and 1001 becomes 10010. Succeeding clock pulses can shift the binary number farther —100100, 1001000, 10010000, etc. Such shifting is, of course, a portion of the multiplication process.

Converters

Rather than use the natural binary number, such as 1101011 (decimal 107), it is usual to go to a system known as binary coded decimal (BCD) with each decimal position taken as a four-bit binary representation. The BCD equivalent of 107 is 0001 0000 0111.

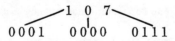

Conversion from the four-bit binary uses a BCD-to-decimal decoder having 10 gates and a decimal-to-bar encoder. The BCD-decimal decoder uses both flip-flop outputs—that is, the positive 1

142

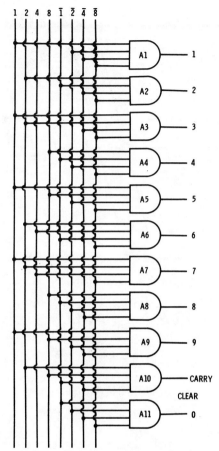

Fig. 12-14. BCD-to-decimal decoder circuit.

output of one inverter and the 0 output of the other inverter. In the circuit of Fig. 12-14, the outputs of these flip-flops are the inputs indicated as 1, 2, 4, 8, $\overline{1}$, $\overline{2}$, $\overline{4}$, and $\overline{8}$ (the bar above a number indicating the inverted input). Then with the flip-flop series indicating 0001 the AND gate A1 fed by 1, $\overline{2}$, $\overline{4}$, and $\overline{8}$ passes current to output (1). That is, 1, $\overline{2}$, $\overline{4}$, and $\overline{8}$ are all at the 1 state and will open gate A1. Fed by 1, 2, 4, and $\overline{8}$, gate A7 opens only when the flip-flops are indicating the binary 0111. Gate A10 produces a negative pulse to reset FF1, FF2, FF3, and FF4 and a positive CARRY pulse to the next higher series of flip-flops.

Actual display of the decimal numerals is usually by use of light-emitting diodes (LEDs). Seven LEDs are arranged as indicated by Fig. 12-15 and commonly known as a 7-bar display. Other parts

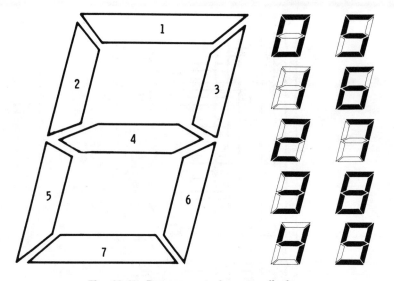

Fig. 12-15. Bar-segment character display.

of this figure show the combinations of these 7 bars necessary to display the ten decimal numerals. It then becomes a matter of designing a matrix (coding network) feeding outputs of the BCD-to-decimal decoder to the proper LED segments (Fig. 12-16).

COMMON USES

While the initial purpose and usage of digital circuits was in computers and calculators, the range of use now extends into just about every phase of life. Digital circuits monitor or aid in maintaining vital life functions of the human body and precisely process food preparation. Presently this book allows for only the description of a digital watch or clock.

As might be expected, digital circuits have been miniaturized through integrated circuits. Fig. 12-17 shows the IC chip for a digital watch and its mounting within the watch interior. Note the four LED display units across the center of the watch. A block diagram of this digital watch circuit is given by Fig. 12-18. The oscillator section uses a quartz crystal with a natural frequency of about 2 MHz (actually 2.097152 MHz) as compromise between size, cost, and frequency stability in an inverter type circuit. Additional inverters provide gain and shaping of pulses. Twenty-one flip-flops in the frequency divider reduce the rate to one pps. Another six flip-flops, with an inverted feedback to the third flip-flop, provides a binary ratio of N/60 and an output of one pulse per minute. Ap-

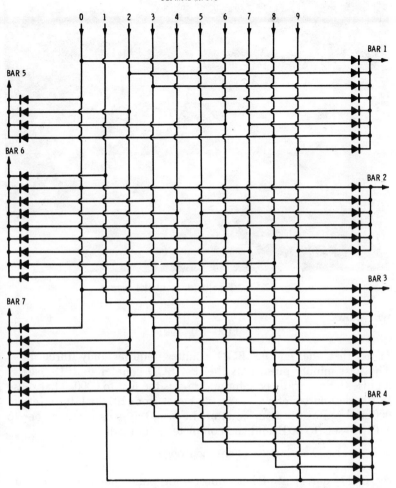

Fig. 12-16. Decimal-to-bar segment encoder matrix.

plication of these pulses is to a half adder and a BCD system of flip-flops acting as a read-only-memory (ROM). The first-four flip-flop memory and associated BCD-to-decimal decoder works through the ten digits in the usual manner. The second set of flip-flops—three in number—needs only to add up to 0110 (6) since there is only sixty minutes in an hour. Actually the BCD 0110 0000 (60) triggers a re-set and puts a positive pulse into the hour counter. Four flip-flops in the hour counter go to 1101 (13) before re-setting to 0001 (1)—never going to a 0000 count as there is no 0 hour in the normal time notation. When the time is 12:59, the BCD counters

145

Fig. 12-17. IC chip for a digital watch.

will show

$$1100 \; 101 \; 1001$$

(note that the middle BCD counter requires only three bits)
The next minute pulse puts the first BCD counter to 1010, carries
a pulse to the middle counter and re-sets the first to 0000. That pulse
to the middle counter changes its count from 101 to 110, re-sets to
000 and pulses the hour counter. In turn, the hour counter goes to
1101, re-sets to 0001 and the complete BCD reads

$$0001 \; 000 \; 0000$$

the BCD equivalent of 1:00. Succeeding pulses put the BCD count
to 0001 000 0001 (1:01), 0001 000 0010 (1:02) . . . 0001 101 1001
(1:59), 0010 000 0000 (2:00), 0010 000 0001 (2:01). . . .

It almost seems necessary to apologize for not covering the digital
circuit subject more thoroughly but that is probably not really possible for any one book. While the calculators, computers, and other
devices are being miniaturized through large-scale integration (LSI)
—advanced integrated circuits—the usage and theory is becoming
larger. Some of this LSI includes the microcomputer IC shown by
Fig. 12-19. Nicknamed "a computer on a chip," the microcomputer
is the heart of the small electronic calculator and includes a read-
only-memory (ROM), a random-access-memory (RAM), an oscillator clock, control decode (programming), a logic unit (ALU),

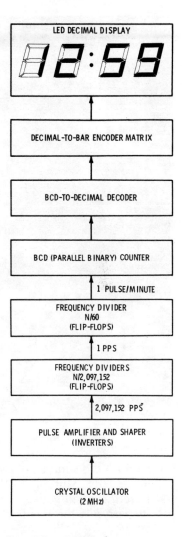

Fig. 12-18. Block diagram of digital watch circuitry.

and an output decode. The photograph in Fig. 12-20 show a comparison of "computer on a chip" with the size of contact lenses.

REVIEW QUESTIONS

1. In the decimal number 9357, what is the least significant bit? What is the most significant bit?
2. How many digits are used in the binary number system? What quantity is indicated by the fourth bit of a binary number? The third bit? What is the decimal equivalent of the binary 1101?

Fig. 12-19. Microcomputer chip.

3. Perform the addition of 0101 plus 1001. The subtraction of 1111 minus 0011. Multiply 0101 times 0011.
4. Describe a half adder. If its inputs are 1 and 0, what is its sum? Its carry? If the inputs are 1 and 1, what is the sum and the carry?
5. What is an AND gate? In an AND gate has inputs of 1, 1, 1, 1, and 0, will its output go to a 1-state level?
6. What is an OR gate? What is the OR gate output level if its inputs are 0, 0, 1, 0, 0, 0, and 0?
7. If an exclusive OR gate has signals A and B (both at the 1-state level) presented at the same instant, what is its output? If A is 1 and B is 0?
8. What is a flip-flop? Describe the basic form of a flip-flop circuit. How many input pulses are required to put the flip-flop through a complete on/off cycle? What is the natural binary

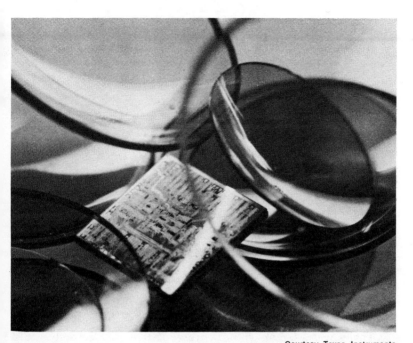

Fig. 12-20. Comparing the size of a "computer on a chip" to contact lenses.

ratio of a series of three flip-flops? If an input signal has a rate of 1000 pulses per second, what is the output rate of a series of three flip-flops?
9. What is BCD? What function does the BCD-to-decimal decoder perform? The decimal-to-bar encoder?
10. Discuss briefly the operation of a digital watch.

Answers to Review Questions

Chapter 1

1. Thales was a Greek who discovered that static electricity could be produced by rubbing a rod of amber.
2. Elektron.
3. Electricity.
4. Molecule and atoms.
5. 104
6. Copper. Carbon and hydrogen. Carbon, hydrogen and chlorine.
7. Electrons, protons, and neutrons.
8. They are repelled from each other. Attracted to each other.
9. Negatively charged.
10. Those electrons beyond the second orbit of an atom that are more or less free to escape from the attraction of its protons.

Chapter 2

1. Static electricity.
2. An insulator is a material that will not pass electrical current. A conductor is a material that passes electrical current readily.
3. Static electricity.
4. Lightning is the movement of static electricity from cloud to cloud or from cloud to earth.
5. Dynamic electricity.
6. No, not a definite link.
7. The N pole.
8. In lines.

9. The copper wire pushes the magnetic lines aside until they reach their limit of elasticity and break or are cut.
10. S-pole.
11. Magnetic induction involves the use of magnetic lines of force to produce electron movement (current).
12. Yes.
13. Begins at zero, increases to a positive maximum, falls back to a zero value, increases to a maximum negative value, and again drops to a zero value.
14. Frequency is the term denoting the number of cycles occurring in one unit of time.
15. Zero.
16. 60 cycles per second, or 60 hertz.
17. Audio frequencies are those creating sound. Radio frequencies are those producing radiating electromagnetic waves.
18. A commutator is a mechanical device for changing alternating current to pulsating direct current.
19. A 3-phase power consists of three separate currents or voltages, each of which is out-of-phase with the others.
20. The transformer has two windings with a common magnetic path. 360 volts.

Chapter 3

1. The atoms within copper are loose and part freely, while those in carbon are closely fitted and have electrons that are tightly held to the nucleus.
2. Conductance.
3. Volt.
4. One ampere.
5. Potential difference is the voltage between two points of a circuit.
6. 60 watts.
7. 14,500 ohms.
8. 3 ohms.
9. Because its electrons tend to move toward that point which is deficient of electrons or positive.
10. A dielectric is an insulating material between charged points or plates.
11. One. 2.9 6.0
12. A capacitor is two plates separated by a dielectric capable of acquiring or repelling a fixed number of electrons when acted upon by a given emf.
13. Inertia is the mechanical property of any body to resist a change in movement.
14. Yes.

15. The electron orbits within the iron line up in the same plane as the coiled wire.
16. Self-induced emf is that force that tends to keep electron flow, or current, moving within a conductor.
17. Inductance is the physical ability to produce self-induction.
18. Shape, size, and concentration of conductor as well as the magnetic characteristics of the core material near the conductor or inside the coiled conductor.
19. The henry.
20. Because it opposes that emf causing any change in the rate of current flow.
21. No.
22. Yes, the magnetic lines of force developed by current through the first inductance acts upon the turns of the second inductance. Therefore, a change in the current through the one inductance can, by way of a common magnetic field, produce a current within the second inductance.
23. The maximum value of the force occurs at a time differing from the time when the maximum movement occurs.
24. In phase.
25. Lag.
26. $\frac{1}{4}$-cycle or 90°.
27. Inductive reactance is the opposition to current through an inductance. $X_L = 6.28fL$.
28. Capacitive reactance is the opposition to current through a capacitor. $X_C = 1/6.28fC$.
29. Impedance is the effective or resultant of all forms of opposition to current. Resistance, inductive reactance, and capacitive reactance.
30. Series resonance is that condition of a circuit having resistance, capacitance, and inductance but minimum impedance since the reactances are equal and opposite in effect. One.
31. The tank circuit has an inductance in parallel with a capacitance and is in resonance. High.

Chapter 5

1. Bell Telephone Laboratories. They also found methods of obtaining germanium and silicon in a nearly pure state.
2. The crystal has a fixed pattern of construction.
3. N-type. A hole is a space unfilled by an electron. A crystal with an excess of holes.
4. P-type.
5. The concentration of impurity in the diffused crystal material is not constant but gradually decreases, or increases, from one side of the crystal to the other.

6. A semiconductor is any device or substance that conducts electrical current readily in one direction while opposing current flow in the opposite direction. The junction semiconductor consists of a section of p-type crystal joined to a section of n-type crystal.
7. Yes. The reversed-bias condition has the p-type section connected to the negative terminal of a battery.
8. That region near the pn junction having a lack of holes and electrons.

Chapter 6

1. The point-contact diode consists of a crystal block mounted on a metal plate, with a sharpened point contact with the opposite side of the crystal. The junction-type diode has a p-type and an n-type crystal bonded or grown together.
2. Yes. Excellent.
3. Zener effect.
4. The silicon controlled rectifier is one of a group of diodes known as thyristors that can be switched between the ON and OFF conducting states. Four. The gate is the control element.
5. Its negative-resistance characteristics.
6. Because small voltage rectification occurs when the tunnel diode is reverse biased.
7. A capacitance.
8. Decrease.
9. For luminescent display of letters and numerals.

Chapter 7

1. The bipolar depends upon the interaction of both electrons and holes, while the unipolar is dependent upon only one type charge carrier. Emitter, base, and collector.
2. The collector current, the base current, and the emitter current. 1.02 mA.
3. Dots of impurity are placed upon opposite sides of the base wafer and heated to alloy the impurity into the wafer. An increased frequency range.
4. Homotaxial is basically an alloy-junction device, while the epitaxial is a grown-junction device.
5. Common-base, common-emitter and common-collector. The common-emitter. Yes.
6. Transfer characteristics and collector characteristics. Alpha or the ratio of collector current to emitter current. Beta or the ratio of collector current to base current.
7. Source, gate, and drain.

8. Its input impedance was severely affected by the polarity of its gate potential. Its gate is insulated from the channel. The depletion type passes drain current regardless of the gate polarity while the enhancement type passes drain current only when the gate-channel junction is forward biased.
9. No. Two.
10. Common-gate, common-source, and common-drain.
11. The forward transconductance of an FET is the ratio of its drain current to its gate voltage. 50 volts.

Chapter 8

1. Discrete components are separate, individual components. Integrated circuits combine the many components into a single unit—usually on a silicon chip.
2. A solid-state resistor consists of a strip of p-type silicon of a given resistivity arranged between metal contacts. A solid-state capacitor has an n-type section as one plate, and a metal plate with a silicon-oxide dielectric.
3. Reduce number of leads.
   No inductors.
   Reduce total capacitance.
   Reduce total resistance.
   Use matching components.
   Substitute transistors and diodes for larger area components.
   Use built-in voltage references.
4. Any generator or other device having a fixed level of current flow regardless of its load. Current sinks pass a given rate of current.
5. A voltage reference provides a constant voltage independent of the supply voltage.
6. Due to direct coupling. Level shifter.
7. It draws little or no current when there is no signal applied.
8. Input level, input impedance, output level, output impedance, power dissipation, supply current, supply voltage, and temperature.
9. No.

Chapter 9

1. To increase the amplitude of a voltage or a current.
2. The base-emitter current.
3. Increases or decreases in base-emitter current produce larger increases or decreases in the emitter-collector current, which in turn increases or decreases an output voltage. 20.

4. A Darlington pair has the emitter circuit of one transistor directly coupled in series with a second transistor. 2500.
5. 30.

Chapter 10

1. Op amps are very-high gain dc amplifiers.
2. Yes. Versatility and reliability.
3. Common-mode rejection is the amplification of only that portion of a signal not common to both IC inputs. Medical and audio equipment.
4. Amplification of the difference between two signals. No.
5. Output 1 has an inverted and amplified ac while output 2 has an amplified ac.
6. An inverter shifts the signal by 180°. A Darlington pair has the emitter circuit of one transistor directly coupled in series with the base-emitter circuit of a second transistor.
7. Performance of mathematical operations.
8. The open-loop gain is the overall gain of an op amp. The closed-loop gain is that of an op amp with negative feedback.
9. 150 ohms and 3000 ohms.
10. In the form of input and feedback impedances.
11. Negative-feedback circuit. A triangular shaped wave. A cosine wave.
12. The differential of a wave is the ratio of changes occurring on the wave. Small blips or spikes of voltage.
13. 501.

Chapter 11

1. Maxwell.
2. 1896.
3. The telephone receiver is placed so that a portion of its output is fed back into the transmitter.
4. The basic principle of an oscillator is an amplifier with a portion of its output fed back to its input. The fed-back current or voltage is that current or voltage fed from the amplifier output to the input to sustain oscillations. Proper amplitude and phase.
5. An rf power amplifier.
6. The Hartley.
7. By changing either the inductance or the capacitance of the resonant "tank" circuit.
8. No.
9. Wavelength is the distance traveled by an alternating current wave during the time of one cycle.
10. 186,000 miles or 3,000,000,000 meters per second.

11. A reflected wave is that current wave caused by a self-induced emf.
12. Radiocommunications are less reliable at frequencies above 30 megahertz since such waves are not usually refracted back from the outer atmosphere.
13. Vertical.
14. Modulation is the addition of intelligence to a radiated wave. Amplitude modulation varies the amplitude of the radiated wave, while frequency modulation varies the frequency of the radiated wave.

Chapter 12

1. 7. 9.
2. Two. 8s. 4s. 13.
3. 1110 (14). 1100 (12). 1111 (15).
4. The half adder uses two AND gates, one OR gate and an inverter to determine states of binary bits to be added and develop the correct SUM and CARRY. 1. 0. The sum is 0 and the carry is 1.
5. An AND gate has an output of 1 only when all of its inputs are at the 1 level. No.
6. The output of an OR gate goes to the 1 level when any one of its inputs is at the 1 level. 1.
7. 0. 1.
8. A flip-flop is any electrical or mechanical device that is stable in either of two conditions. A flip-flop is basically two inverters with the output of the second feeding the input of the first. Two. N/8. 125.
9. Binary coded decimal. The BCD-to-decimal decoder translates the binary number to the electrical equivalent of the decimal number. The decimal-to-bar encoder selects the LED segments necessary for proper display of the decimal number.
10. A digital watch uses a chain of flip-flops to derive properly spaced pulses that drive counters and display elements.

Index